RESEARCH REQUIRED TO SUPPORT COMPREHENSIVE NUCLEAR TEST BAN TREATY MONITORING

Panel on Basic Research Requirements in Support of
Comprehensive Test Ban Monitoring

Committee on Seismology
Board on Earth Sciences and Resources
Commission on Geosciences, Environment, and Resources
National Research Council

National Academy Press
Washington, D. C. 1997

NOTICE: The project that is the subject of this report was approved by the Governing Board of the National Research Council, whose members are drawn from the councils of the National Academy of Sciences, the National Academy of Engineering, and the Institute of Medicine. The members of the committee responsible for the report were chosen for their special competences and with regard for appropriate balance.

This report has been reviewed by a group other than the authors according to procedures approved by a Report Review Committee consisting of members of the National Academy of Sciences, the National Academy of Engineering, and the Institute of Medicine.

This study was supported by Award No. PO-950011 between the National Academy of Sciences and the Air Force Office of Scientific Research and Grant No. EAR-9526501 between the National Academy of Sciences and the Advanced Research Projects Agency (via the National Science Foundation). Any opinions, findings, conclusions, or recommendations expressed in this publication are those of the author(s) and do not necessarily reflect the view of the organizations or agencies that provided support for this project.

Library of Congress Catalog Card Number 97-68150
International Standard Book Number 0-309-05826-0

Additional copies of this report are available from:

National Academy Press
2101 Constitution Ave., NW
Box 285
Washington, DC 20055
800-624-6242
202-334-3313 (in the Washington Metropolitan Area)
http://www.nap.edu

The cover art was created by Carrie Mallory.

The cover is an oil painting of a large valley near Wolfville, Nova Scotia.

Ms. Mallory received her Bachelor of Fine Arts from the Cooper Union. She frequently exhibits at juried shows in Northern Virginia. She takes many of her themes from nature and has provided art for a number of NRC reports.

Printed in the United States of America

PANEL ON BASIC RESEARCH REQUIREMENTS IN SUPPORT OF COMPREHENSIVE TEST BAN MONITORING

THORNE LAY, *Chair*, University of California, Santa Cruz
SUSAN L. BECK, University of Arizona, Tucson
ALFRED BEDARD, Environmental Technology Laboratory, Boulder, Colorado
ADAM M. DZIEWONSKI, Harvard University, Cambridge, Massachusetts
JOHN R. FILSON, U.S. Geological Survey, Reston, Virginia
WILLARD J. HANNON, JR., Lawrence Livermore National Laboratory, Livermore, California
DONALD V. HELMBERGER, California Institute of Technology, Pasadena
WILLIAM A. JESTER II, The Pennsylvania State University, University Park
SHELDON LANDSBERGER, University of Illinois, Urbana
PETER MIKHALEVSKY, Science Applications International Corporation, McLean, Virginia
JOHN A. ORCUTT, University of California, La Jolla
PAUL G. RICHARDS, Lamont-Doherty Earth Observatory, Palisades, New York
ROBERT C. SPINDEL, University of Washington, Seattle
BRIAN STUMP, Los Alamos National Laboratory, New Mexico
RODNEY W. WHITAKER, Los Alamos National Laboratory, New Mexico

Staff

CHARLES MEADE, Study Director
THOMAS M. USSELMAN, Senior Staff Officer
VERNA J. BOWEN, Administrative Assistant
JUDITH L. ESTEP, Administrative Assistant

COMMITTEE ON SEISMOLOGY

THOMAS H. JORDAN, *Chair*, Massachusetts Institute of Technology, Cambridge
RALPH ARCHULETA, University of California, Santa Barbara *(appointed 3/20/97)*
SUSAN BECK, University of Arizona, Tucson *(appointed, 3/20/97)*
STEVEN M. DAY, San Diego State University, California *(term ended 12/31/96)*
THOMAS C. HANKS, U.S. Geological Survey, Menlo Park, California
CHARLES A. LANGSTON, The Pennsylvania State University, University Park *(term ended 12/31/96)*
THORNE LAY, University of California, Santa Cruz
STEWART A. LEVIN, Mobil Exploration & Production Technical Center, Dallas, Texas
STEPHEN D. MALONE, University of Washington, Seattle
T. GUY MASTERS, University of California, San Diego *(appointed 3/20/97)*
JAMES R. RICE, Harvard University, Cambridge, Massachusetts
PAUL G. SOMERVILLE, Woodward-Clyde Consultants, Pasadena, California
ANNE M. TREHU, Oregon State University, Corvallis *(term ended 12/31/96)*
JOHN E. VIDALE, University of California, Los Angeles

Staff

CHARLES MEADE, Senior Staff Officer
VERNA J. BOWEN, Administrative Assistant
JUDITH L. ESTEP, Administrative Assistant

COMMISSION ON GEOSCIENCES, ENVIRONMENT, AND RESOURCES

GEORGE M. HORNBERGER, *Chairman*, University of Virginia, Charlottesville
PATRICK R. ATKINS, Aluminum Company of America, Pittsburgh, Pennsylvania
JAMES P. BRUCE, Canadian Climate Program Board, Ottawa, Ontario
WILLIAM L. FISHER, University of Texas, Austin
JERRY F. FRANKLIN, University of Washington, Seattle
THOMAS E. GRAEDEL, Yale University, New Haven, Connecticut
DEBRA KNOPMAN, Progressive Foundation, Washington, D.C.
KAI N. LEE, Williams College, Williamstown, Massachusetts
PERRY L. MCCARTY, Stanford University, California
JUDITH E. MCDOWELL, Woods Hole Oceanographic Institution, Massachusetts
RICHARD A. MESERVE, Covington & Burling, Washington, D.C.
S. GEORGE PHILANDER, Princeton University, New Jersey
RAYMOND A. PRICE, Queen's University at Kingston, Ontario
THOMAS C. SCHELLING, University of Maryland, College Park
ELLEN SILBERGELD, University of Maryland Medical School, Baltimore
VICTORIA J. TSCHINKEL, Landers and Parsons, Tallahassee, Florida
E-AN ZEN, University of Maryland, College Park

Staff

STEPHEN RATTIEN, Executive Director
GREGORY SYMMES, Assistant Executive Director
JEANETTE SPOON, Administrative Officer
SANDI FITZPATRICK, Administrative Associate
MARQUITA SMITH, Administrative Assistant/Technology Analyst

The National Academy of Sciences is a private, nonprofit, self-perpetuating society of distinguished scholars engaged in scientific and engineering research, dedicated to the furtherance of science and technology and to their use for the general welfare. Upon the authority of the charter granted to it by the Congress in 1863, the Academy has a mandate that requires it to advise the federal government on scientific and technical matters. Dr. Bruce Alberts is president of the National Academy of Sciences.

The National Academy of Engineering was established in 1964, under the charter of the National Academy of Sciences, as a parallel organization of outstanding engineers. It is autonomous in its administration and in the selection of its members, sharing with the National Academy of Sciences the responsibility for advising the federal government. The National Academy of Engineering also sponsors engineering programs aimed at meeting national needs, encourages education and research, and recognizes the superior achievements of engineers. Dr. William A. Wulf is president of the National Academy of Engineering.

The Institute of Medicine was established in 1970 by the National Academy of Sciences to secure the services of eminent members of appropriate professions in the examination of policy matters pertaining to the health of the public. The Institute acts under the responsibility given to the National Academy of Sciences by its congressional charter to be an adviser to the federal government and, upon its own initiative, to identify issues of medical care, research, and education. Dr. Kenneth I. Shine is president of the Institute of Medicine.

The National Research Council was organized by the National Academy of Sciences in 1916 to associate the broad community of science and technology with the Academy's purposes of furthering knowledge and advising the federal government. Functioning in accordance with general policies determined by the Academy, the Council has become the principal operating agency of both the National Academy of Sciences and the National Academy of Engineering in providing services to the government, the public, and the scientific and engineering communities. The Council is administered jointly by both Academies and the Institute of Medicine. Dr. Bruce M. Alberts and Dr. William A. Wulf are chairman and vice chairman, respectively, of the National Research Council.

Preface

In 1995, the Air Force Office of Scientific Research, and the Air Force Phillips Laboratory requested a review of their seismic research programs in support of nuclear test verification efforts. For this task, the National Research Council's Committee on Seismology appointed a panel of 10 seismologists. At the group's first meeting, however, the sponsors described Department of Defense (DoD) proposals to eliminate the Air Force programs and to consolidate all of the DoD research efforts related to nuclear test monitoring. As described in this report, these proposals were approved and eventually implemented in fiscal year 1997, raising obvious difficulties for the work of the panel.

The organizational changes at DoD occurred while the Comprehensive Nuclear Test Ban Treaty (CTBT) was under negotiation in Geneva. To ensure compliance with the CTBT's ban on nuclear explosions, a global seismic, hydroacoustic, infrasonic, and radionuclide data collection system is to be deployed as part of an International Monitoring System (IMS). The United States has indicated that it would monitor international treaty compliance using these unclassified IMS data, together with additional National Technical Means (NTM). To meet this challenge, the newly created Nuclear Treaty Program Office (NTPO) within DoD plans to broaden the support for research in the Defense Department's treaty monitoring efforts. Recognizing that the National Research Council had already formed a panel with significant seismological expertise on nuclear monitoring, NTPO requested modifications to the ongoing Air Force study to consider a broader range of disciplines and research needs for CTBT verification. In response to this request, six panelists were added, two each from the fields of hydroacoustics, infrasound, and radionuclide monitoring, and the scope of the study was enlarged (see Appendix A).

In all, the original and enlarged panel met five times over a period of 14 months between November 1995 and January 1997. In the course of its work, the panel received briefings from representatives of the following offices and agencies: Air Force Office of Scientific Research, Air Force Phillips Laboratory, Nuclear Treaty Program Office, Center for Monitoring Research, Office of Nonproliferation and National Security (Department of Energy), Lawrence Livermore National Laboratory, Los Alamos National Laboratory, Pacific Northwest National Laboratory, Air Resources Laboratory (National Oceanic and Atmospheric Administration), and the Air Force Technical Applications Center (AFTAC). One of these briefings (AFTAC) was presented at the Secret level. In addition, some members of the panel attended classified on-site meetings at AFTAC to discuss national monitoring operations and research needs in the fields of seismology, hydroacoustics, and radionuclides. The panel also received a tour and briefing on the operations at the Center for Monitoring Research. Throughout the study, the panel received valuable assistance from its liaison representatives: Stanley Dickinson (AFOSR), James Lewkowicz (Phillips Laboratory), Ralph Alewine (NTPO), Steve Bratt (NTPO), and David Russell[†] (AFTAC).

In response to its charge, the panel's report describes the research needs and associated infrastructure needed to promote high-confidence monitoring of the CTBT by the United States. For this work, the panel relied on its expertise in the fields of seismology, hydroacoustics, infrasound, and radionuclides to analyze the role of each discipline alone and in conjunction with others in specific treaty monitoring capabilities. The report concludes that continued basic research will improve these capabilities, effectively lowering the threshold for CTBT compliance and eventually achieving U.S. monitoring goals. Developing synergies between monitoring technologies is important in this effort, but doing so will require a significant research program because data sets for most of the monitoring technologies have not been available in the past for small events in regions of interest. Throughout the report, the panel notes mechanisms to transition research results to monitoring operations. Such efforts will be essential for future improvements in monitoring capability.

[†] David Russell was a member of the original AFORSR-PL review panel. His participation was changed to liaison when the charge was enlarged by NTPO.

Contents

EXECUTIVE SUMMARY 1

1 INTRODUCTION: THE COMPREHENSIVE NUCLEAR TEST BAN TREATY 7
- 1.1 Nuclear Testing Treaties, 7
- 1.2 Requirement of Safeguards, 9
- 1.3 Role of Science in Nuclear Test Treaty Monitoring, 9
- 1.4 Monitoring Compliance with the CTBT, 14
- 1.5 The International Monitoring System, 16
- 1.6 U.S. Operations, Research, and Development Structures, 20
- 1.7 Transitions in the U.S. Research Program, 21

2 CTBT MONITORING TECHNICAL CHALLENGES THAT DRIVE RESEARCH 23
- Introduction, 23
- 2.1 Physical Phenomena: Source Excitation, Signal Propagation, and Recording, 24
- 2.2 Functions of Monitoring Systems, 30
- 2.3 Monitoring Infrastructure, 46

3 MONITORING TECHNOLOGIES: RESEARCH PRIORITIES 49
- Introduction, 49
- 3.1 Seismology, 49
- 3.2 Hydroacoustics, 60
- 3.3 Infrasonics, 65
- 3.4 Radionuclides, 71
- 3.5 Other Technologies, 78
- 3.6 Opportunities for New Monitoring Synergies, 79
- 3.7 On-Site Inspection Methods, 81

4 U.S. RESEARCH INFRASTRUCTURE 83
Introduction, 83
4.1 Structure of Current DoD and DOE Programs, 83
4.2 Research Program Balance, 85
4.3 Coordination with Other National and International Efforts, 86
4.4 Requirements for Long-Term Stability and Effectiveness, 87

5 CONCLUSIONS AND RECOMMENDATIONS 89
5.1 CTBT Monitoring Challenges, 89
5.2 Recommendations, 90

REFERENCES 97

APPENDIXES
A STATEMENT OF TASK 101
B RESEARCH SUPPORT HISTORY 103
C SEISMIC EVENT LOCATION 107
D SEISMIC MAGNITUDES AND SOURCE STRENGTHS 113
E HYDROACOUSTICS 121
F INFRASONICS 125
G RADIONUCLIDE SOURCE TERM RANGES FOR DIFFERENT TEST SCENARIOS 131
H ACRONYMS 137

Executive Summary

Each State Party undertakes not to carry out any nuclear weapon test explosion or any other nuclear explosion, and to prohibit and prevent any such nuclear explosion at any place under its jurisdiction or control. -- Article 1.1, Comprehensive Nuclear Test-Ban Treaty

On September 24, 1996, President Clinton signed the Comprehensive Nuclear Test Ban Treaty (CTBT) at the United Nations Headquarters. Over the next five months, 141 nations, including the four other nuclear weapons states,[1] added their signatures to this total ban on nuclear explosions. By the Law of Treaties, the signatories are bound to abide by the provisions of the CTBT prior to its entry-into-force, effectively creating an immediate moratorium on nuclear weapons testing. Formally, the treaty will enter into force 180 days after instruments of ratification have been deposited by the 44 States with nuclear power reactors listed in the CTBT (but no earlier than September 24, 1998). Notably, this list includes three countries that have not, as of April 1997, signed the treaty: India, Pakistan, and North Korea. As of February 18, 1997, two nations have ratified the treaty.

To help achieve verification of compliance with its provisions, the treaty specifies an extensive International Monitoring System (IMS) of seismic, hydroacoustic, infrasonic, and radionuclide sensors. The IMS will be developed under the guidance of a Preparatory Commission that has already been established and begun to meet. On request, the system will provide monitoring data to each State Party for use in its national treaty verification efforts. In association with the Conference on Disarmament, a prototype IMS has been operating continuously since January 1, 1995. Analysis of these data, using state-of-the-art scientific and technical expertise across a wide range of disciplines will play a critical role in the immediate effort to monitor and verify the comprehensive ban on nuclear explosions.

To sustain and advance these CTBT monitoring capabilities, there is a need for a strong basic research effort to support improved analysis of IMS and National Technical Means (NTM) data. Recognizing this challenge, the Air Force Office of Scientific Research, Air Force Phillips Laboratory, and the Nuclear Treaty Program Office (Office of the

[1] Russia, China, France, and the United Kingdom.

Secretary of Defense) requested a panel of the National Research Council's Committee on Seismology to identify the broad range of basic research issues that would strengthen national capabilities to monitor a global ban on nuclear explosions. The charge for this study is presented in Appendix A. In response, this report describes specific research activities in the fields of seismology, hydroacoustics, infrasound, and radionuclide monitoring. The panel concludes that research in all of the monitoring technologies is needed if the U.S. monitoring system (when fully deployed) is to achieve stated U.S. monitoring goals. To help shape a national research effort in these areas, the report also describes current research programs that support nuclear monitoring, and it recommends strategies to increase their effectiveness and stability.

Monitoring CTBT compliance will be more challenging than prior nuclear testing treaties because it will require high confidence identification of any nuclear explosions in the atmosphere, underwater, underground or in space amidst a significant background of earthquakes, volcanic eruptions, storms, meteor impacts, and conventional explosions such as mining blasts. President Clinton recognized this challenge and the limitation of current monitoring capabilities when he committed the United States to an absolute ban of nuclear explosions. In defining national safeguards for the CTBT, he stated, "I recognize that our present monitoring systems will not detect with high confidence very low yield tests. Therefore I am committed to pursuing a comprehensive research and development program to improve our treaty monitoring capabilities and operations."[2]

Successful use of the IMS and NTM data will rely on the following elements; understanding of source excitations, accounting for signal propagation or advection through the Earth, automated recording and telemetry by instrumentation, and analysis of the signals for event detection, location, and identification. Basic research has enhanced the performance of each of these elements, but extensive new work is needed to meet the technical challenges of monitoring a comprehensive ban on nuclear explosions, especially because there has been limited experience using synergistic monitoring strategies and in-country (regional) data. This effort will require a global system to detect the distinctive seismic and acoustic waves, radioactive materials, and radiation emanating from nuclear explosions. Through the analysis of signals that have passed through the complex Earth system, scientific and technical expertise will play a critical role in the identification of nuclear events amidst the significant background noise of weather phenomena, earthquakes, and conventional explosions.

In support of these efforts, there is a need for a broad program of basic research in the fields of seismology, hydroacoustics, infrasound, and radionuclide monitoring. In the context of this report, basic research means long-term research on fundamental issues associated with CTBT monitoring technologies, as distinct from communications and computer systems development, instrumentation engineering, and software automation. The panel emphasizes that it is important to buffer these basic research efforts from short-term operational needs, otherwise creativity and innovation will be curbed and the long-term benefit to CTBT monitoring will be diminished. Priority research areas for each discipline are described below. Additional topics are listed under the section of Research Synergy.

SEISMOLOGY

For decades, nuclear testing treaties have been verified using seismic monitoring of teleseismic signals (i.e., signals that are recorded more than 2000 km from the source). Teleseismic signals are weak, however, for the small events that will be of interest for CTBT monitoring. Consequently, treaty verification will necessitate increased dependence on "regional" signals (i.e., signals that are measured at distances significantly less than 2000 km). Pushing the seismic monitoring threshold downward to include precise event locations and high confidence identification for small events at regional distances is the primary motivation for continued seismological research. To support this effort, a prioritized list of research topics include:

1) Improved characterization and modeling of regional seismic wave propagation in diverse regions of the world.
2) Improved capabilities to detect, locate, and identify small events using sparsely distributed seismic arrays.

[2] White House Press Release, August 11, 1995.

3) Theoretical and observational investigations of the full range of seismic sources.
4) Development of high-resolution velocity models for regions of monitoring concern.

HYDROACOUSTICS

Monitoring sound waves in the oceans is a well-advanced discipline, primarily as a result of investments in Anti-Submarine warfare. To date, however, there has been relatively little research on the use of hydroacoustic signals to monitor underground and atmospheric explosions. Given that the proposed IMS hydroacoustic network will use a small number of sensors, with no directional capabilities, the panel concludes that the system will have extremely limited detection and location capabilities. Because of these deficiencies, there is a need for research on synthesizing hydroacoustic data with seismic, infrasonic, and radionuclide information and to assess the capability of the integrated system to monitor within national goals. As part of this effort, prioritized research topics include:

1) Improvements in source excitation theory for diverse ocean environments, particularly for earthquakes and for acoustic sources in shallow coastal waters and low altitude environments.
2) Understanding the regional variability of hydroacoustic wave propagation in oceans and coastal waters and the capability of the IMS hydroacoustic system to detect these signals.
3) Improved characterization of the acoustic background in diverse ocean environments.
4) Improving the ability to use the sparse IMS network for event detection, location, and identification and developing algorithms for automated operation.

INFRASOUND

At present, the U.S. has only a few experts in infrasound, and virtually no infrastructure for research in atmospheric monitoring using low-frequency sound waves. Thus, the primary research issues associated with CTBT monitoring involve first-order questions regarding the characterization, understanding, and reduction of background noise phenomena. To this end, prioritized research activities include.

1) Characterizing the global infrasound background using the new IMS network data.
2) Enhancing the capability to locate events using infrasound data.
3) Improving the design of sensors and arrays to reduce noise.
4) Analyzing signals from historical monitoring efforts.

RADIONUCLIDES

A wide range of research is needed to strengthen the capabilities of radionuclide monitoring, to reduce the time delay between potential explosions and radionuclide detection, and to facilitate the work of On-Site Inspection teams. Prioritized research topics include:

1) Research to improve models for backtracking and forecasting the air borne transport of radionuclide particulates and gases.
2) Research and data survey to improve the understanding of source term data.[3]
3) Understanding of atmospheric rain-out and underground absorption of radionuclides from nuclear explosions.
4) Assessment of the detection capabilities of the IMS radionuclide network.
5) Research on rapid radiochemical analysis of filter papers.
6) Development of a high resolution, high efficiency gamma detector capable of stable ambient temperature monitoring.

RESEARCH SYNERGY

Effective verification will require operation of the monitoring technologies, NTM, and intelligence assets as an integrated system. As indicated above, there are great opportunities to utilize synergies between the different CTBT monitoring technologies because energy propagating in the Earth

[3] Source terms refer to the amounts of diagnostic radionuclides likely to be released by explosions of different sizes in diverse environments. See Appendix G.

system can couple from one medium to another (air ↔ water, air ↔ land, or land ↔ water). In this area, it is particularly important to investigate synergies to enhance the performance of the IMS hydroacoustic network. Priority research topics include:

1) Improved understanding of the coupling between hydroacoustic signals and ocean island-recorded T-phases, with particular application to event location in oceanic environments.
2) Integration of hydroacoustic, infrasound and seismic wave arrivals into association and location procedures.
3) Use of seismo-acoustic signals together with an absence of radionuclide signals for the identification of mining explosions.
4) Explore the synergy between infrasound, NTM, and radionuclide monitoring for detecting, locating, and identifying evasion attempts in broad ocean areas.
5) Determine the false alarm rate for each monitoring technology when operated alone and in conjunction with other technologies.

DATA ACCESS

The panel strongly recommends that scientific researchers have near real-time access to the IMS data streams received by the U.S. National Data Center (NDC). The research facilitated by this access will provide broad-based quality control and allow the development of monitoring algorithms using actual monitoring data. In turn, the use of the data by the broad research community will enhance the U.S. monitoring capability, increase confidence in the operation of the monitoring system, and serve as an increased deterrent to potential evaders. A previous NRC report (NRC, 1995) detailed the benefits to CTBT monitoring from open distribution and multiuse of seismic data under the Group of Scientific Experts Technical Test - 3 experiment.[4] The current panel strongly endorses the conclusions and recommendations on open data access in NRC (1995), emphasizing that they should be applied to all IMS technologies and to the operation of IDC. As this report was written, the panel was aware that the policies governing access to IMS data were still undefined. In the panel's view, the most effective strategies for improving U.S. monitoring capabilities will facilitate research contributions from the broadest segments of the scientific community by allowing open distribution and multiuse of IMS data streams. To facilitate this research, the panel strongly recommends that the U.S. Government should formulate a policy supporting open distribution of IMS data for scientific research.

RESEARCH FUNDING AND PROGRAM BALANCE

Substantially increased funding and closer agency coordination will be required to pursue the panel's recommendations on research needs and to support U.S. efforts to meet national monitoring goals. The appropriate funding level for basic and applied research in universities and private industry must be established apart from that required to develop IMS systems, which is an applied technology area. The panel concludes that it is important to stabilize the budgets for the CTBT research program, with a multi-year commitment firmly establishing the viability of this research area for intellectual resources in universities and private companies. This stability is essential for training technically competent scientists and researchers who will participate in U.S. monitoring operations. Without it, bright young researchers will not enter the fields supportive of CTBT monitoring.

OPPORTUNITIES FOR TECHNOLOGY TRANSFER

Finally, the panel concludes that increased numbers of Ph.D. level research staff at the U.S. National Data Center would help to promote technology transfer to the operational regime. Technical training and sophistication is essential for recognizing and rapidly incorporating research advances into operational systems. The panel also recommends the establishment of a CTBT research test bed as an additional means to transition research efforts. This would require an accessible system that replicates significant aspects of the

[4] The Group of Scientific Experts Technical Test –3 is an ongoing demonstration test of the operational capabilities for the existing seismic stations in the IMS.

IMS and U.S. NDC monitoring system, with real-time data processing capabilities and archives of historical data.

Much of this report discusses basic research in seismology, hydroacoustics, infrasound, and radionuclide monitoring needed to support enhanced CTBT monitoring. In Chapter 1, the panel frames these issues with a discussion of the technical challenges associated with monitoring the CTBT, the importance of the Presidential safeguards for monitoring, and the role of the IMS and basic research in achieving these goals. The chapter also describes current programs for basic research in support of nuclear monitoring. Chapter 2 outlines the basic monitoring procedures that drive the most important research problems. Chapter 3 summarizes the research issues, and discusses strategies to enhance monitoring capabilities through basic and applied research. Chapter 4 describes past and present programs to support these research efforts. The chapter also discusses the characteristics of an effective long-term research program and the mechanisms to transition the research results to an operational environment. Following the conclusions and recommendations (Chapter 5), the report includes several appendices with more detailed discussions of the research issues and monitoring challenges for the fields of seismology, hydroacoustics, infrasound, and radionuclides. Finally, Appendix H defines the many acronyms that are used in this report.

1

Introduction: The Comprehensive Nuclear Test Ban Treaty

After a half century in which nuclear weapons were developed, tested, and used, a Comprehensive Nuclear Test Ban Treaty (CTBT) banning all nuclear explosions has been negotiated and signed by 142 countries (as of February 18, 1997) including the United States. Although the U.S. Senate has yet to give advice and consent to U.S. ratification of the CTBT (two nations had ratified the CTBT as of February 18, 1997), it appears likely that nuclear explosion testing is over after a history of more than 2090 explosions. (The July 29, 1996, underground explosion in China may have been the last nuclear test.) Verification of compliance with the CTBT will be a major concern of many nations in both the short and the long term, and it requires a vigorous research program to enhance capabilities to identify violations, minimize false alarms, and thus maintain confidence in compliance.

This chapter discusses the technical challenges of the CTBT in the context of previous nuclear arms control treaties. It then describes the importance of the Presidential safeguards for monitoring and the contributions of the International Monitoring System (IMS) and basic research toward achieving these goals. Finally, the chapter describes current programs for basic research in support of nuclear monitoring.

1.1 NUCLEAR TESTING TREATIES

The Limited Test Ban Treaty (LTBT) of 1963 prohibited explosions in the oceans, atmosphere, and space by the signatories, bringing to an end the perils of radioactive fallout from testing. However, the LTBT did not ban underground nuclear explosions. Significant new nuclear weapons development and underground testing took place in the ensuing decades. To monitor this treaty, the United States used a combination of atmospheric infrasound, seismic, hydroacoustic, radionuclide, and satellite methods to ensure that explosions in banned environments could be detected. However, given that testing could take place underground, these monitoring efforts were limited in scope. At the same time, it was recognized that seismological monitoring of underground nuclear testing in other countries provided a means by which to monitor advances in weapons technologies reflected in their underground testing practices. Following the recommendations of the Berkner panel (Berkner et al., 1959), the United States deployed unclassified and classified seismological recording systems to enhance the national capability to detect, identify, and characterize underground nuclear explosions. Since 1959, the Department of Defense (DoD) has sustained a seismological research program to support the analysis of these data, recognizing the central role that this discipline has played in monitoring the development of nuclear weapons, as well as in monitoring for treaty compliance.

The bilateral 1974 Threshold Test Ban Treaty (TTBT) placed an upper limit on the yield of U.S. and Soviet underground nuclear explosions equivalent to 150 kilotons (kt) of TNT. The monitoring challenge of this treaty was to estimate accurately the yield of the largest Soviet underground nuclear explosions. The seismic magnitude of these explosions was approximately 6.1. There was little difficulty in locating or identifying such large events because they produced detectable signals over the entire surface of the Earth. (The seismic wave amplitudes from a 150 kt event are about 50 times larger than those from a 1 kt explosion.) Because seismological monitoring was the primary means to verify the TTBT, there was a vigorous research program to address the question of yield estimation using seismic data. Progress in this area eventually enabled accurate yield estimates. It also documented significant systematic variability of seismic magnitudes at fixed yields arising from variability in wave transmission properties near the major U.S. and Soviet test sites.

The CTBT prohibits all nuclear explosions, effectively extending the LTBT to include underground tests. The 90-page text of the CTBT is about 50 times longer than the text of the LTBT, in large part because of extensive provisions (in the CTBT) for verification. The formal treatment of verification issues in the CTBT will continue to be developed and documented in extensive detail over the next few years. The treaty and its Protocol mention six different Operational Manuals, not yet written, that will spell out the technical and operational requirements:

1. Seismological Monitoring and the International Exchange of Seismological Data;
2. Radionuclide Monitoring and the International Exchange of Radionuclide Data;
3. Hydroacoustic Monitoring and the International Exchange of Hydroacoustic Data;
4. Infrasound Monitoring and the International Exchange of Infrasound Data;
5. the International Data Center; and
6. On-Site Inspections.

The CTBT and its Protocol specify an IMS consisting of 170 seismic stations, 80 stations monitoring relevant airborne radionuclides, 11 hydroacoustic and T-phase[5] monitoring stations, and 60 infrasound stations, with associated global communications and integrated signal processing infrastructure; an International Data Center (IDC) to collect, archive, process, and distribute data and processing products; as well as procedures for On-Site Inspection (OSI). Importantly, however, the responsibility for determining treaty compliance rests with the States Parties, not with the CTBT organization. Thus, the United States and other nations can use IMS data, along with any additional sources of objective information available to them, to monitor the treaty. However, if a suspect event occurs, an OSI request must be based on information collected by the IMS, on any relevant information obtained by national technical means (NTM) in a manner consistent with generally recognized principles of international law, or on a combination of these. Monitoring the CTBT is anticipated to entail a long-term effort, in which it is desirable to keep costs as low as possible while achieving monitoring objectives. One of the primary purposes of this report is to assess what research activities will be required by the United States to ensure effective monitoring of the CTBT. A previous report of the National Research Council (NRC, 1995) addressed ways in which the CTBT seismic monitoring effort can contribute to independent areas of national concern, such as earthquake monitoring and basic research on the Earth system. IMS data from other monitoring technologies similarly have potential multiple use for research and Earth system monitoring, and ensuring that these data are generally available is important. An extensive history of CTBT negotiations is presented in Pounds (1994) and in several papers in Husebye and Dainty (1996).

---

[5] In this report, T-phase refers to (1) a seismic signal that originates from converted hydroacoustic energy at the ocean-Earth interface or, conversely, (2) a hydroacoustic signal arising from a converted elastic wave at the ocean-Earth interface. Examples of T-phases include hydroacoustic recordings of earthquakes and seismic recordings of suboceanic explosions. The IMS will exploit the properties of T-phases by using ocean-island seismic stations to augment the hydroacoustic network. These seismic stations are termed "T-phase" stations.

1.2 REQUIREMENT OF SAFEGUARDS

The CTBT is a zero-yield treaty, meaning that no nuclear explosions are allowed, including any that might be deemed "Peaceful Nuclear Explosions" (PNE's) under the 1976 Peaceful Nuclear Explosions Treaty. When President Clinton announced that the U.S. negotiating policy for the CTBT would adopt the zero-yield decision, he recognized that this places great demands on any treaty monitoring system and stated, "I recognize that our present monitoring systems will not detect with high confidence very low-yield tests. Therefore, I am committed to pursuing a comprehensive research and development program to improve our treaty monitoring capabilities and operations."[6] In fact, no viable system can monitor compliance with the CTBT with high confidence to the smallest possible yields, because explosions with a nuclear yield below a few pounds of TNT equivalent can be carried out, and it is technically possible to make such small explosions undetectable by the standard monitoring techniques. Given this reality, the issue of CTBT verification capability involves a political determination of the risk that is acceptable as a function of the yield level, the confidence level that is desired, the resources that are available; and other factors.

It has long been recognized that verification capability for a CTBT is set by overall political agendas, which differ widely between countries. In that setting, this report addresses the ability of technical systems to monitor at various levels and the research that is required to achieve and enhance these abilities. In general, the technical systems put in place to monitor the CTBT will be under pressure to detect, locate, and identify small "events" underground, underwater, and in the atmosphere with high confidence and accuracy. This pressure translates into requirements for research to constantly improve the capabilities and results from technical monitoring systems in a cost-effective way.

The United States has specified precise monitoring levels for international CTBT compliance. The monitoring levels are not geographically uniform, and the specifics are classified. For the purpose of this report, an appropriate distillation of the President's requirements for CTBT monitoring is the phrase "a few kilotons evasively tested" in selected areas of the world (see, for example, the U.S. working paper for the Conference on Disarmament of May 1994 [United States, 1994]). Evasive testing involves any of a number of scenarios for masking or muting the signals from a nuclear explosion. For underground testing, detonation in a large cavity can reduce the magnitude of seismic signals significantly. This is of concern in limited geographic areas of the world, and there are limited opportunities for evasive testing in the oceanic and atmospheric environments. The Presidential Safeguard calling for sustained research and development efforts in support of CTBT monitoring recognizes that evasion scenarios provide strong motivation to continue research in monitoring technologies.

Although any nuclear explosion that is detected will be of interest to the United States and will be a violation of the CTBT if a signatory nation is involved, specification of a target threshold for monitoring guides the assessment of research priorities. In this report the panel focuses on U.S. national needs because international requirements are not defined (in general, based on negotiation history, the United States appears to have more stringent CTBT monitoring requirements than most other countries, although all nations that sign the treaty intend for there to be a total ban on testing). Although some estimates of the potential monitoring capability of the IMS suggest a high-confidence threshold of around 1 kt, nonevasively tested, on a global basis, this is not a formal design criterion. The network design was constrained strongly by considerations of cost and uniformity of coverage. Because the system is not yet fully deployed, validation of the projected monitoring capability will take several years, and even if a fully coupled 1 kt level is achieved, it will not satisfy the U.S. monitoring objective for areas in which evasion is considered viable. The U.S. NTM will augment IMS capabilities, even with the enhancement of the latter as research progresses.

1.3 ROLE OF SCIENCE IN NUCLEAR TEST TREATY MONITORING

Throughout the past four decades the United States has monitored foreign nuclear testing in

[6] White House Press Release, August 11, 1995.

order to assess the associated programs of nuclear weapons development, as well as monitor compliance with previous test ban treaties. The present situation, with a signed CTBT, is different in that the primary driver is monitoring compliance with an arms control treaty that bans *all* nuclear explosions, and the goals of detecting, identifying, and characterizing frequent (large) nuclear explosion signals are absent. Even more so than in the past, solid technical grounds are imperative for CTBT monitoring, because the lower signal-to-noise ratios inherent in low-yield monitoring and the plethora of nonnuclear events with signals similar to nuclear explosions make it far more difficult to monitor this treaty than previous test bans.

The IMS technologies will provide data for identifying a probable nuclear explosion if the signals can be distinguished from natural phenomena or mining explosions. Additional technologies such as satellite radiation sensors will play a key role in monitoring the atmosphere and space environments. In addition, satellite imaging capabilities can be used to detect testing operations such as drilling or device emplacement, along with sensing environmental changes associated with a nuclear test (ground disruption, crater formation, test analysis facilities).

Any CTBT monitoring system will have practical limits in terms of the capabilities of the system to detect, locate, and identify events. These limits are imposed both by cost considerations that constrain data acquisition and processing and by intrinsic constraints of monitoring technologies. A complete interpretation of monitoring limits must allow for the possibility of various evasion approaches, such as muffling a nuclear explosion signal by detonation of the device in a preexisting cavity (decoupling) or obscuring the explosion signal by simultaneous detonation with an earthquake, quarry blast, or mine collapse. More than 50 years of research underlies the present ability to use the various wavetypes in diverse environments for monitoring applications. The significant progress that has been achieved has provided the technical basis for moving forward with CTBT negotiations. However, national objectives for ensuring international compliance with a total ban on nuclear explosions place extreme demands on all of the monitoring technologies and operational systems, and there is a need for continuing research to enhance the entire U.S. CTBT monitoring system. Furthermore, none of the technologies mentioned above can monitor hydronuclear tests (experiments for which the fissile component of the device is modified to reduce the nuclear yield to levels on the order of a few hundredths of a pound of TNT).

The physical processes associated with nuclear explosions produce distinctive sources of acoustic waves (in the atmosphere or ocean), elastic waves (underground), possible releases of radioactive materials (underground events can result in some immediate release of radioactive gases and particulates or delayed seepage of radioactive gases), or characteristic radiation (in space). These signals and products then propagate through or are advected by the Earth system with various transmission effects and eventually may be detected by different types of sensors placed around the planet's surface or on satellites. The background noise, comprised of signals from nonnuclear events including weather, and/or the physical limitations of the sensors constrain the signals that can be detected and the attributes of the signals that are recorded (e.g., frequency bandwidth, rate of time sampling, background levels of radioactive materials). Signals recorded at different sensors must be retrieved from the field and associated with a common source using general knowledge of how such signals propagate, and the time of origin and location of the source must be estimated. Attributes of the recorded signals, corrected for propagation and instrumentation effects, are then used to identify the type of source, ideally distinguishing nuclear explosion signals from earthquakes or other nonnuclear phenomena. All monitoring technologies share these fundamental elements: source excitation, signal propagation or advection, recording instrumentation, event association, event location, and event identification. They also share the technological challenges of data retrieval and automation of data analysis.

Basic research contributes to monitoring the CTBT by enhancing the performance of monitoring elements, and nuclear test monitoring research programs in the past four decades have carried out research on all of the essential elements mentioned above. The primary technical challenge associated with the CTBT is related to the fact that even very small tests are banned. Signals from small events are more difficult to detect, and the number of earthquakes, chemical explosions, and natural or man-made radioactive sources whose signals have characteristics similar to those of nuclear explosions increases as the size of the events of concern decreases.[7] For buried, well-coupled underground explosions, a 1 kt explosion produces seismic magnitudes of about 3.8-4.5 (depending on the source environment). Thus, CTBT monitoring for an explosion yield threshold of 1 kt for underground explosions that are detonated without special concealment efforts would require monitoring capabilities that provide high-confidence detection and identification of events with seismic magnitudes of 3.8-4.5 or larger on a global basis. The projected capabilities of the IMS are for global seismic magnitude monitoring levels of about 4.0 (for high-confidence event detection and location; reliable identification probably will have a higher threshold).[8] It is possible to reduce the seismic magnitude for a given yield by decoupling, but such evasion efforts also must consider overhead imaging and radionuclide monitoring capabilities.

Decoupling and other evasion scenarios appear to be limited geographically by geological conditions, but in areas of concern (typically involving states with advanced technological capabilities and prior nuclear testing experience), high-confidence detection and identification of events down to seismic magnitudes of about 2.5 is required to monitor fully decoupled 1 kt explosions. Typically, the experience from teleseismic monitoring[9] is that reliable identification of events requires a detection threshold approximately 0.5 magnitude units below the target threshold. This increases the number of events that must be examined and, if possible, identified. Regional seismic signals (involving waves that travel in the crustal waveguide) might exhibit a smaller difference in identification and location thresholds.

U.S. monitoring goals apparently will not be met by the stations of the IMS alone but may be met by the use of additional stations, possibly including mobile monitoring capabilities. To put these numbers in perspective, current global earthquake bulletins produced by the earthquake monitoring community are complete for magnitudes of about 4.5 and larger, and complete catalogs are obtained for magnitudes as small as 2.0 in localized areas with regional earthquake monitoring systems. In this setting, assessments as to whether a given monitoring system is adequate for treaty verification will be driven, in part, by perceptions of the plausibility of treaty evasion.

As the magnitude threshold decreases, the number of detected earthquakes and chemical explosions increases. At the same time, the number of stations in a fixed network that will detect a given event and the distance range at which detections are made decrease. (Wave amplitudes tend to decrease with distance as the energy spreads outward.) Given these factors, the detection, location, and identification of small events by combined IMS and NTM assets involve the analysis of signals recorded at regional distances where wave propagation is often complicated and regionally varying. For example, "regional" seismic waves within 1000 km of the source typically reverberate in the crustal waveguide, potentially obscuring the source type and complicating event location. Even when only a few stations provide data, it is necessary to have a high-confidence location with an area of uncertainty no greater than 1000 km^2. This requirement reflects operational requirements and is mandated by the On-Site Inspections provisions of the treaty (Protocol to the CTBT, Part IIA). The challenge of precisely locating and confidently identifying all small events at some low-magnitude threshold given sparse monitoring networks is formidable. Meeting it requires a sustained basic research program in support of CTBT monitoring.

Current U.S. treaty monitoring capabilities are founded on knowledge and research discoveries that emerged from university research programs over the past five decades. For example, in the realm of seismic monitoring, almost all of the key concepts and methods originated at universities.

[7] Every seismic magnitude unit reduction of the monitoring threshold for a network involves about an order-of-magnitude increase in the number of events that must be included. Scaling relations between explosion yield and seismic magnitude indicate that for every factor of 2 reduction in the yield monitoring threshold, there will be a factor of about 1.7 more natural events that must be identified.

[8] Because the capabilities of the IMS seismic system will not be uniform on a global basis, the actual threshold will be lower than 4.0 in many places. For example, Eurasia will be monitored to levels of approximately 3.5. See figure 4.2.

[9] Teleseismic refers to signals recorded more than 2000 km from the source.

These include seminal research on event detection, location estimation, discrimination, yield estimation, seismic magnitude scales, global velocity models, seismic wave attenuation, seismic coupling, seismic regionalization, seismic wave excitation, and broadband seismic recording systems. Much of this university research was funded by DoD research programs. Given the historical evolution of current monitoring capabilities, it is reasonable to assert that the long-term development of new or enhanced monitoring capabilities is largely dependent upon sustaining a fundamental research program in relevant areas. This is also critical for training the next generation of scientifically capable personnel to support long-term CTBT monitoring operations.

Scientific support of CTBT monitoring is not restricted to fundamental research programs conducted by universities. Other types of scientific research are essential for national monitoring capabilities. These include applied or exploratory developmental research, which also involves the development and testing of new scientific understanding and technologies for the functions of treaty monitoring. For example, a proposed seismic magnitude scale for regional distance observations may require extensive testing and validation prior to acceptance (or rejection) by the monitoring organization. This type of work is usually conducted by private contractors, who pursue sophisticated developmental work on specified problems, often using data (which may be classified) from actual monitoring operations. This activity serves a key role in the transition of scientific advances from universities to government agencies. The technical expertise of these private companies is provided mainly by university graduates who are then trained in test ban monitoring by their work with private contractors.

Even more directly linked to operational needs is the implementation of scientific systems such as communication, computer, and software platforms. A major element of this effort for CTBT monitoring involves automated processing by computer software. This advanced developmental work usually involves state-of-the-art technology, and again private contractors play a major role. Some agencies sustain internal scientific capabilities (e.g., the Department of Energy's National Laboratories) for pursuing a range of fundamental, exploratory developmental, and advanced developmental research, but by their very nature these tend to be mission-oriented programs. Innovative basic research is pursued more easily in a university setting.

Ideally, the national treaty monitoring capability built on this scientific infrastructure should be established prior to a political agreement; however, the political forces driving treaty negotiations are generally not coupled with treaty monitoring capabilities. Although the lag of a monitoring capability may not preclude signing a treaty, it can be a reason for delay in ratification, as in the case of the 1974 TTBT, which was not ratified by the United States until 1990, in part because confidence-building measures that would have calibrated monitoring systems were not set in place until the late 1980s.

When scientific capabilities and understanding evolve in parallel with the political process, there is often confusion or turmoil with respect to the role and adequacy of the scientific contributions. Such was the case for the TTBT, when ongoing research was coming to an understanding of the variations of seismic magnitudes associated with 150 kt explosions in different regions of the world at about the same time (and in some cases even before) that political assertions of treaty violation were being made based on earlier (flawed) scientific understanding. The ultimate validation of the seismological research results brought about by the bilaterally (U.S.-U.S.S.R.) monitored explosions of the Joint Verification Experiment (JVE) in 1988 reinforced the value of seismological monitoring of treaties. This experience emphasized the need for the operational regime to be fully informed and responsive to research advances and to implement them rapidly in operations; for policy makers to be aware of the technical limitations of the monitoring systems; and for the technical community to be aware of the policy applications of its results and the need to be clear about their strengths and limitations.

The fundamental technical bases for most nuclear test monitoring technologies have not changed substantially over the past few decades. As discussed in Chapter 2, a solid foundation of physical principles underlies all of the monitoring technologies anticipated to play key roles for the CTBT. However, every technology has intrinsic capabilities and limitations imposed by the monitoring system, the physical nature of the technology, the number and nature of nonnuclear events with signal characteristics similar to those of nuclear explosions, and the heterogeneity of the Earth. Enhancing the operational capabilities of all disciplines can be achieved by a combination of a full spectrum of coordinated research and development conducted by universities, private companies, and government laboratories and systematic implementation of the methodologies with continuing empirical calibration of each monitoring element. Development of new, unexpected monitoring capabilities also requires a basic research effort. The very nature of the CTBT monitoring environment is that monitoring capabilities steadily will evolve and improve, but they will never achieve 100 per cent confidence in being able to identify and attribute small or evasively conducted nuclear test explosions. It is desirable to establish rapidly the monitoring capabilities of the system when all IMS and NTM assets are in place and empirically calibrated, (a process that will take several years) and then to monitor improvements in these capabilities brought about by continuing research efforts and discoveries. Throughout the monitoring process there will be problem events that defy identification by standard means, and it is important to develop a system that can exploit all available data to resolve the cases that are of concern. Clear communication among the research, operational, and policy arenas about the state of the monitoring capabilities and their continuing evolution must be promoted by the monitoring infrastructure in order to avoid misunderstandings at any level. As the President has stated, the U.S. decision to support the CTBT necessitates a strong commitment to a research program to enhance monitoring capabilities.

The level of effort in basic research that must be sustained involves a policy decision that should be informed by appraisals of the present capabilities of the system, assessments of the potential for improvement, and the benefits of enhanced levels of treaty compliance. For global monitoring systems, this value-added assessment should take into account dual-use benefits between CTBT monitoring and other national interests such as earthquake and volcano hazards, meteorite monitoring, and nuclear reactor emission monitoring. All of these activities will benefit greatly from multiuse of the data collected by the IMS and NTM for CTBT monitoring purposes, and they will build confidence in compliance with the CTBT.

It is important to recognize that the role of science is multifaceted, particularly for monitoring technologies associated with complex Earth systems. For example, the effort to understand seismic or acoustic signals from nuclear explosions should not be conducted in isolation from understanding such signals resulting from earthquakes, quarry blasts, landslides, or other sources of wave motions, some of which have properties similar to those of nuclear explosions. Similarly, one cannot simply study sources in one part of the world and generalize the results to the entire globe. Geological processes have produced great heterogeneity in the planet's interior that influences the propagation of seismic and acoustic energy on each path from source to receiver. Similarly, wind patterns, ocean currents, and ocean floor topography vary from place to place, and human activities are region dependent. There is an intrinsic need to attain an understanding of the source and propagation effects for all monitoring technologies for all significant source types and specific regions of the world.

Awareness of the complex range of scientific issues associated with nuclear test monitoring has led to the establishment of broadly defined research efforts in the past, addressing issues from fundamental source and propagation theory to the calibration of specific paths. In addition to stimulating basic scientific investigations, treaty monitoring efforts motivate many technical research endeavors, such as those leading to automation of signal processing, enhanced telecommunications, and sensor development. It is essential to maintain some separation between immediate operational needs and the basic research effort because focus on the former may not allow one to develop the new perspective that completely alters the context and the panoply of available resources. Availability of new data also results in unexpected research problems that can be solved only by a broadly based program.

Although the balance of fundamental research investigations and applied developmental efforts may be expected to shift somewhat with time as any field matures, the CTBT is a context in which continued basic research will play a key role even as current U.S. monitoring objectives are achieved over the next decade. This importance follows from the very nature of CTBT monitoring, which involves "problem events" that have characteristics similar to those of nuclear explosions. These could be violations of the treaty or "false alarms," politically and economically costly mis-identifications of natural events as presumed nuclear tests (with purported high confidence). There are a vast number of detected events that routine capabilities will dispose of readily, but some problem events will always remain (which has been well established by past monitoring efforts). This is particularly true for the small events important to CTBT monitoring if efforts at evasion are deemed to be of serious concern. By definition, conventional processing fails to give high-confidence identification for problem events, and innovative approaches will be needed to reduce the yield equivalent at which problems arise. The new approaches may come from unexpected directions that must be sustained by basic research endeavors. The panel returns to this issue in Chapter 4.

1.4 MONITORING COMPLIANCE WITH THE CTBT

Given the U.S. national objective of monitoring the CTBT to a level of a few kilotons evasively tested in key areas of the world, it is possible to estimate the operational requirements of the monitoring network based on historical experience with nuclear explosions of known yields. For each monitoring technology a theoretical monitoring threshold can be established, which is defined in terms of a high confidence level (often defined as 90 per cent confidence of recognizing a violation if one is attempted) in detection and identification of all events (as nuclear or non-nuclear) with a certain monitoring characteristic. The relevant characteristics vary among technologies, depending on the particular measurement procedures. For example, in seismic monitoring, many measures involve seismic magnitudes (logarithmic scales based on the amplitude of ground shaking produced by different events). For hydroacoustic signals, the relevant measurements are typically described in terms of logarithms of pressure units such as micropascals (μPa). These measures, corrected for predictable propagation effects such as the decrease of acoustic and seismic wave amplitudes with increasing distance from the source caused by expansion of the wavefront, are indirect measures of the source properties. For example, the precise fraction of nuclear explosion energy or earthquake energy released into the elastic wavefield is not known (it is on the order of at most a few percent) and must be established empirically. Thus, to translate the monitoring capability of seismology into a corresponding bound on nuclear explosion size requires an empirical calibration of magnitude in terms of explosion yield.

Given the past 50 years of nuclear testing there is an empirical basis for relating current-day thresholds for different monitoring systems to equivalent nuclear explosion yields, albeit with some uncertainty. For example, a 1-kilogram explosion (chemical or nuclear) detonated in the SOFAR channel (SOund Fixing and Ranging; see Appendix E) of most regions of the world's deep oceans will be detected by several hydroacoustic stations. A 1 kt explosion in the SOFAR channel in most oceanic regions would drive most of the hydroacoustic instruments around the world off-scale (there are, however, oceanic areas that may be blocked). Thus, the favorable sound transmission properties of the ocean define a low effective detection level, at which the challenge will be in distinguishing conventional explosions (e.g., for seismic exploration) from other events.

The solid Earth is far less efficient in transmitting seismic waves, but there is an extensive basis for relating seismic magnitude to explosion yield for certain test sites. For example, based on announced yields for various test sites around the world, one particular seismic magnitude, m_b(ISC) (the 1-second period P-wave magnitude determined by the International Seismological Centre [ISC]; see Appendix D for a discussion of various magnitudes) has been found in one study to have the following mean values for a 150 kt explosion at each site (Adushkin, 1996):

- 5.74 (Nevada Test Site, tuff);
- 6.12 (Semipalatinsk, East Kazakhstan);
- 5.97 (Novaya Zemlya, Russia);
- 5.93 (Lop Nor, China); and
- 6.04 (Mururoa, French Polynesia).

Such numbers form the basis for seismic monitoring of compliance with the TTBT.

Linear relationships between seismic magnitude and the logarithm of the explosion yield make it possible to determine "yield-scaling" curves that allow prediction of the yield, and its uncertainty, for any given magnitude. For a 1 kt explosion, such relations for m_b(ISC) predict corresponding magnitudes of

- 3.87 (Nevada Test Site, tuff);
- 4.45 (Semipalatinsk, East Kazakhstan);
- 4.32 (Novaya Zemlya, Russia);
- 4.3 (Lop Nor, China); and
- 4.5 (Mururoa, French Polynesia).

These magnitudes have higher uncertainty than the 150 kt values above, because 1 kt is near or below the level of the smallest events that were available for calibrating the yield-scaling curves. Also, the effort over the past two decades was focused on the 150 kt level associated with TTBT compliance, and the uncertainty increases as the yields vary from this value.

Such data provide a basis for translating a monitoring system capability defined in terms of seismic magnitude threshold into a corresponding yield threshold. The magnitudes given above typically correspond to the central point of the distribution for a given yield, and somewhat lower values (by about 0.2 magnitude units) are needed to provide a 90 per cent confidence level at 1 kt.[10] Thus, the above values suggest that teleseismically detecting and identifying 90 per cent of the 1 kt nuclear explosions near the Chinese test site would require a seismic monitoring system with capabilities to detect and identify all events above m_b(ISC) of 4.3-0.2 = 4.1. Typically, however, teleseismic identification requires magnitudes about 0.5 units above the detection and location threshold,[11] thereby lowering the detection and location threshold for the system to 3.6. Evasive decoupling would lower the seismic detection level of the monitoring system by another 1.5-1.85 magnitude units (factors of 30 to 70 reductions in amplitudes for a given yield.[12]) Thus, the experience from teleseismic monitoring suggests that a detection and location capability near magnitude 1.8-2.1 would be required to confidently identify an evasively tested 1 kt explosion near the Chinese test site..[13]

At these low magnitudes, however, monitoring will be done at regional distances. Importantly, the experience with regional signals suggests that the difference between the thresholds for detection or location and identification will be less than 0.5 (as in the teleseismic case). In the best case, this offset could be zero, and the detection requirement would be increased to approximately 2.3-2.6.

In this way, monitoring goals expressed in terms of yield levels and assumptions about decoupling capability are translated into monitoring capabilities that a seismometer network and an associated data center must be designed to handle. An obvious complication is that there are large (0.7 magnitude units, corresponding to factor of 5 seismic ground motion amplitude variations) systematic differences in magnitudes expected for a 1 kt explosion, and this variation is found even among the only five calibrated sites mentioned above. The reason is that variations in coupling of energy from the explosion into the seismic signals (caused largely by rock strength and porosity at each site) and variations in seismic wave attenuation properties (associated with crustal and upper-mantle tectonic history and thermal structure) affect the observed seismic motions. Thus, a globally uniform seismic magnitude monitoring capability implies a nonuniform explosion yield monitoring capability. Fortunately, an understanding of the reasons for differences in magnitude levels between calibrated test sites gives us some predictive capabilities. For example, the highly attenuating structure of the tectonically active western United

[10] There is uncertainty in all yield-scaling relationships with seismic magnitude. To ensure that 90% of the events with a given yield are observed, one must consider events $1.28 \, x \, \sigma_{mb}$ magnitude units below the mean magnitude for that yield, where σ_{mb} is the uncertainty in the yield-scaling for that region. At low yields near 1 kt, the uncertainty in yield for a given magnitude is about a factor of 2, and $1.28 \, x \, \sigma_{mb}$ is about 0.2 magnitude units.

[11] Teleseismic identification generally requires more information than detection or location.

[12] Decoupling is discussed further in Section 3.2. The decoupling factor itself appears to vary from about 70, for full decoupling, down to around 10 for some partially decoupled U.S. and Soviet shots in which the cavity was too small to achieve the greatest possible effect. A factor of 30-70 is probably appropriate for purposes of planning the monitoring network—translating to a reduction on logarithmic magnitude scales by 1.5 to 1.85 units.

[13] In practice, satellite monitoring may offset some of the concern about decoupling for a specific location.

States is mainly responsible for the relatively low seismic magnitudes for Nevada Test Site (NTS) explosions, and other events in the same region experience similar systematic reductions in seismic wave amplitudes for a given source energy level. One can predict, with reasonable confidence, that other tectonically active areas around the world will have similar magnitude reductions relative to geologically stable areas.

To generalize the above discussion, a rational assessment of monitoring capability for the CTBT will require three steps. The first involves an assessment of the intrinsic monitoring capability in terms of signals acquired for each of the monitoring technologies. Second, there is a need to assess the opportunities to merge the data streams from different technologies to enhance detection and identification capabilities. The third step involves an empirical translation of all monitoring measurements into a corresponding explosion yield monitoring threshold, false-alarm rate, and uncertainty. With this background, a policy decision can be made about the performance of the monitoring system and the level of compliance assurance. If it is not sufficient, additional assets must be deployed or research carried out to bring the level down sufficiently.

1.5 THE INTERNATIONAL MONITORING SYSTEM

The CTBT includes specific plans for an extensive IMS that will collect, process, and distribute seismic, hydroacoustic, infrasonic, and radionuclide data collected from global arrays of sensors. The seismic, hydroacoustic, and infrasonic data will be transmitted continuously (or archived at the site and available by dial-up request) to an International Data Center (IDC). The mission of the IDC will be "to support the verification responsibilities of States Parties to the CTBT by providing objective, scientifically-demanding and technically-demanding products and services necessary for effective global monitoring." This will involve the automated and interactive combination of data ("fusion") from different monitoring technologies and the production of a list of event locations (without specific identification), compiled in the IDC Reviewed Event Bulletin (REB). This bulletin, along with raw and processed data, will be provided by the IDC to National Data Centers (NDCs) of the States Parties, which can in turn incorporate them into their national treaty monitoring activities. In addition, the IDC may screen events and carry out screening operations upon request of and for a State Party.

The U.S. NDC will be operated by the Air Force Technical Applications Center (AFTAC), which has historically had the lead U.S. role in monitoring foreign nuclear weapons testing. During the Group of Scientific Experts Technical Test 3 (GSETT-3,[14]) the U.S. NDC considered three mechanisms to provide the scientific research community with open access to all IMS seismic data:

1) an interactive World Wide Web data request procedure;
2) continuous telemetry of windowed seismogram segments to the U.S. Geological Survey (USGS) National Earthquake Information Service (NEIS) for earthquake monitoring applications; and
3) continuous telemetry of all IMS seismic signals to the Data Management Center (DMC) of the Incorporated Research Institutions for Seismology (IRIS) for incorporation into the IRIS data archives, which are the primary data source for U.S. seismological researchers.

In practice, only limited amounts of IMS data (primarily from U.S. stations) have been transmitted by the above mechanisms during GSETT-3. In part these data transfers were limited by logistical problems of establishing the communication infrastructure and protocols from the U.S. NDC. Open distribution of the IMS data by these mechanisms has also been limited, however, by the lack of clear policy on access to international IMS data.[15] As this report was being written, the panel was aware that such policies are still undefined and that they may only be resolved through international discussions at the Preparatory Commission. Recognizing that this is an issue of

[14] GSETT-3 is an ongoing demonstration test of the operational capabilities for the existing seismic stations in the IMS.

[15] Currently, the U.S. NDC provides international IMS data to DoD and DOE supported CTBT researchers as needed for individual projects. DoD has indicated that this policy will continue in the future.

continuing discussion and negotiation in the U.S. Government and at the Preparatory Commission, the panel notes that the treaty contains a strong statement supporting open access to IMS data for scientific research and that there is no language in the treaty suggesting data restrictions. Specifically,

> "The provisions of this Treaty shall not be interpreted as restricting the international exchange of data for scientific purposes." (Article IV, paragraph 10)

A previous NRC report (NRC, 1995) detailed the benefits to CTBT monitoring from multiuse of seismic data streams. The report concluded that open access to IMS seismic data streams would encourage the dual use of such data for earthquake and CTBT research and that it would lead to mutual and unpredictable benefits to both fields. The current panel endorses the data access recommendations from NRC (1995), emphasizing that they should be applied to all IMS technologies. In the panel's view, the benefits of basic research to CTBT monitoring will be severely limited if IMS data is only provided to a limited group of U.S. researchers supported by DoD and DOE CTBT programs. In the panel's view, the most effective strategies for improving U.S. monitoring capabilities will facilitate research contributions from the broadest segments of the scientific community by allowing open access and multiuse of IMS data streams. To facilitate this research, the panel strongly recommends that the US Government should formulate a policy supporting open distribution of IMS data for scientific research.

For the purposes of this report, the panel discusses and recommends research in seismology, hydroacoustics, infrasound, and radionuclide monitoring assuming that there will open access to all of the IMS data for scientific research. Recognizing that this issue has not been resolved, the panel justifies its approach in two ways. First, by discussing the benefits to monitoring from multiuse of IMS data, the panel intends to contribute to policy debate on this issue. Second, the panel's charge from NTPO clearly anticipates the importance of the IMS data as a foundation for a strong research program. Specifically,

> What are the basic research problems remaining in the fields of seismology, hydroacoustics, infrasonics and radionuclides that should be pursued to meet national and international requirements for nuclear monitoring? The panels work on this question should anticipate quality of data to be made available in the future, *in particular those data from the CTBT International Monitoring System.* (emphasis added)

The monitoring stations of the IMS are specified in Annex 1 to the Protocol to the CTBT. The seismic monitoring network will involve 50 stations comprising a *Primary Network*, all providing continuous seismic recordings to the IDC via satellite and telephone communications systems. About 30 of the Primary stations will involve small-aperture, vertical-component high-frequency seismic arrays, and all stations will have a three-component broadband seismic recording system. The IMS seismic system will also include 120 *Auxiliary Network* stations, all with three-component sensors. Only two of these will include arrays. Data from the Auxiliary Network will be available for dial-up or Internet access from the IDC, as well as being maintained in archives at the responsible NDCs for special data requests. Telemetry of selected Auxiliary stations (usually those closest to an initial event location) will be requested by the IDC according to criteria established during the automated event location and processing steps. This achieves significant economy relative to continuous telemetry of all of the data but limits the role of Auxiliary Network data in the initial definition and association of events. The on-demand status of Auxiliary Network data implies that its most important use will be identification, with lesser applications for location and detection. Because of its comparatively lower use, Auxiliary Network data will not be calibrated to the same degree as Primary Network data, vis-à-vis detection and identification, except for stations processed for national purposes. The Auxiliary Network does provide a backup to the Primary Network should data flow be interrupted.

The planned locations of the International Seismic Monitoring Stations are shown in Figure 1.1A. The network is far from being fully deployed (Figure 1.1B), but it is more complete and has a longer operational history than any other IMS technology. The prototype IDC, which has operated since January 1, 1995, was receiving data from 35 (14 arrays, 21 three-component) Primary and 51 (15 of which are used infrequently because of poor cost-benefit and/or performance) Auxiliary stations as of January 1997. The primary function of this seismic system is to monitor underground explosions, but as discussed later, ground motion recordings may also provide information about explosions in the water and atmosphere. Figure 1.1A makes it clear that the political or technical process of CTBT negotiations brought about a relatively uniform global distribution of seismic stations, but even then, there are large regions of the continents and oceans that have few seismic stations. The complementary monitoring capabilities provided by different sensors and NTM must be assessed in these areas, and if adequate monitoring thresholds to meet U.S. CTBT monitoring goals for specific areas are not achieved, additional NTM assets may be deployed. This is true for all of the monitoring systems.

A new 60-station infrasound network is also being deployed for the IMS, involving sensitive atmospheric pressure gauges that can detect acoustic waves in the frequency band excited by atmospheric nuclear explosions. This system primarily will monitor atmospheric events but may also have a role in monitoring underground and underwater environments. The distribution of IMS infrasound stations is shown in Figure 1.1A. In a large number of cases the stations will be near seismic or hydroacoustic stations, which will allow comparison of signals recorded by the different sensors. There has been no extensive infrasound network in operation since the early 1960s, and even given the historical record from infrasound stations that recorded many atmospheric explosions in the 1950s, it is likely that further operational experience will be needed to understand all sources of background atmospheric sounds that must be discriminated from the signals from small explosions. Data from this network will provide valuable information about phenomena such as volcanic eruptions, meteors, and microbaroms (sounds associated with ocean waves). The prototype IDC was receiving data from four infrasound stations as of January 1997, with many stations yet to be deployed.

The oceanic environment will be monitored in part by the IMS hydroacoustic network of 11 systems: 6 single hydrophone sensors (underwater pressure gauges that detect pressure waves in the ocean) and 5 island seismic stations for recording seismic T-phases. The sparseness of the hydroacoustic network (shown in Figure 1.1A) stems from several factors:

1) the political resolution of several States Parties not to include sophisticated hydroacoustic monitoring systems in the IMS,
2) the high cost of hydrophone installations,
3) the efficient sound transmission properties of the oceans (which enable low monitoring thresholds at least in the deep oceans), and
4) the concentration of land-based sensors in the northern hemisphere.

The IMS hydrophone systems will lack the capabilities of hydrophone arrays, which are able to track underwater moving objects using array processing methods. This component of the IMS data collection effort is not exploiting state-of-the-art capabilities as a result of policy decisions and the interplay between seismic and hydroacoustic monitoring capabilities. Additional U.S. hydroacoustic systems (arrays) may provide extended capabilities for U.S. monitoring purposes, although the long-term status of those systems is highly uncertain. The availability of global, unclassified hydroacoustic data offers great research potential for studying underwater phenomena, such as volcanic eruptions, seaslides and turbidity currents, and submarine earthquakes, so there are again many dual-use applications of the data from this system. The prototype IDC was receiving data from four hydroacoustic stations and one T-phase seismic station as of January 1997.

The seismic, infrasonic, and hydroacoustic systems have similar requirements for data transmission and processing, and standardized formats, communications protocols, and parallel processing algorithms have been established for them by the IDC. The details of processing differ because of the distinctive noise conditions; for example, because of the low signal-to-noise ratios characteristic of infrasound signals, detection of infrasonic signals requires correlation detectors rather than ratios of short-term signal power to long-term signal power as used for seismic waves. The basic product of the analysis of each type of signal is an event bulletin, which will include events uniquely detected by one class of signals as well

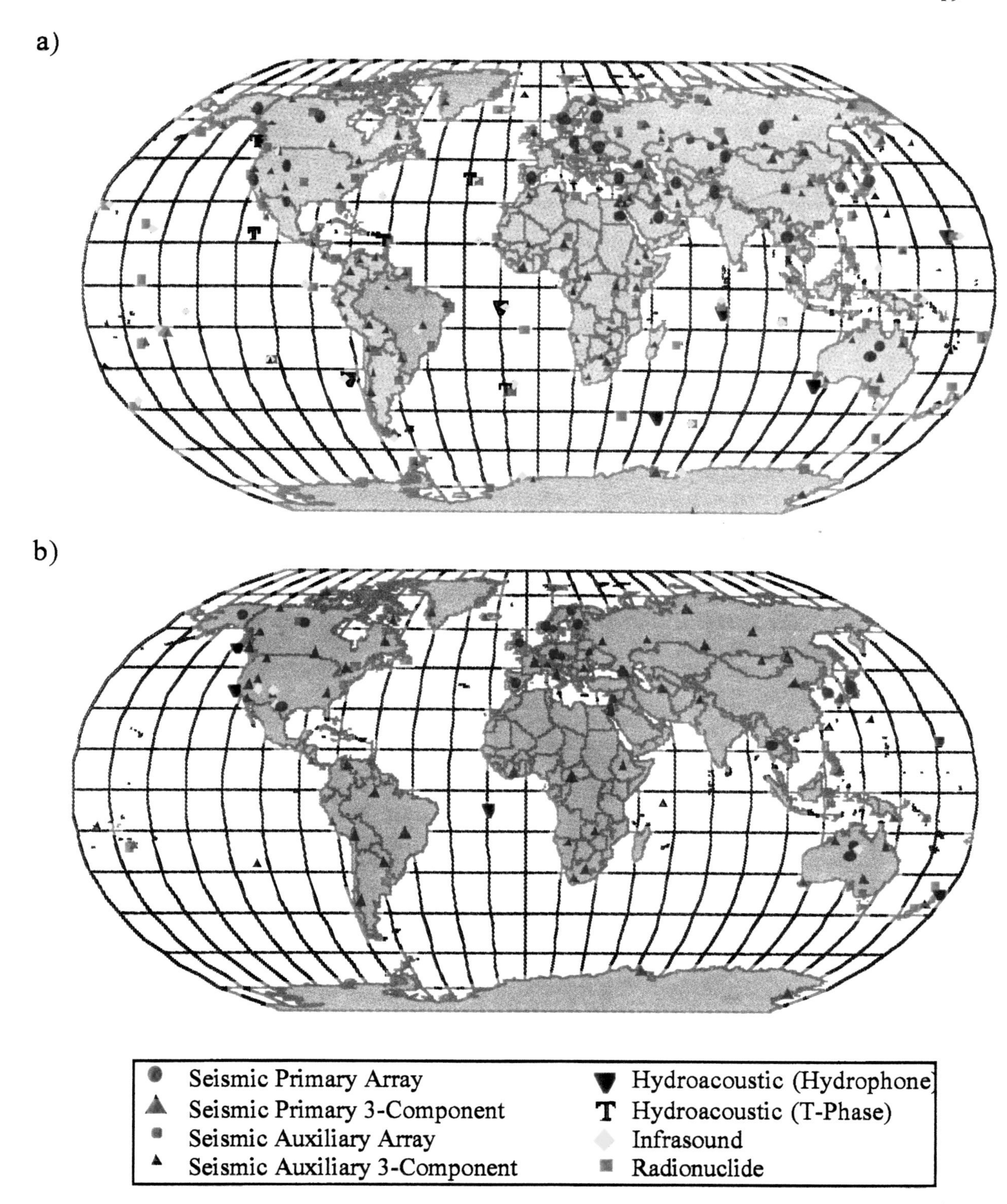

FIGURE 1.1A) Stations of the planned International Monitoring System. B) Stations of the International Monitoring System in operation as of March 21, 1997.

as events detected by a combination of wavetypes from different technologies.

There will be 80 radionuclide stations in the IMS, all of which involve surface stations that sample large volumes of air. All of these systems will collect particulates, and 40 stations will include noble gas detectors. High-resolution gamma-ray spectra will be obtained daily from the particulate and gas samples, and these spectra will be transmitted to the IDC. The CTBT protocol identifies 16 radionuclide laboratories that will maintain specific analytic capabilities for confirming the presence of nuclear debris on the original samples when the gamma-ray spectra indicate radionuclide anomalies. The prototype IDC is presently using a four-level reporting system for the radionuclide methodology: Level 1—natural radionuclides within normal station observations, Level 2—natural radionuclides outside of normal station levels, Level 3—fission or activation products within normal station observations, Level 4—fission products outside normal station observations. A Level 4 report will be more comprehensive than the other reports, and all anomalous fission or activation products will be indicated in the Fission Product Event Bulletin. This report will trigger the use of atmospheric transport models to backtrack the radionuclides' source location. As of January 1997, 21 radionuclide stations were providing data to the prototype IDC.

Some of the stations currently providing data to the prototype IDC are temporary and will be replaced when final IMS stations are installed, and some will require upgrade of current capabilities before meeting IMS standards. However, the current stations are providing a realistic test of data flow for the projected IMS. When the IMS is fully deployed, it will produce a data stream of about 10 gigabytes per day flowing into the IDC. The complete data set acquired by the IDC will be transmitted to the U.S. NDC for the purpose of CTBT monitoring. There will be similar analysis for the detection of events conducted by both the IDC and the U.S. NDC; however, NDC operations will differ in two keys ways. First, the U.S. NDC will incorporate additional data streams from national seismic, hydroacoustic, infrasonic, radionuclide, and satellite sensors, as well as other NTM. Second, it will attempt to identify each event at or above specified monitoring thresholds to ensure that no nuclear tests have taken place. A list of suspect events, along with a bulletin of all events that have been analyzed, will be communicated to the U.S. government. The latter list and at least some of the additional NTM used by the U.S. NDC will not be available routinely to the research community, although under certain conditions, such as the case of problem events, classified research activities may be conducted by the research community using the restricted data. If the IMS data were available to the research community, the future CTBT research program would involve extensive research on many of the actual monitoring signals. This would greatly enhance the impact of unclassified research on the monitoring regime and enable applications of treaty monitoring data to other areas of national concern (such as hazard monitoring).

1.6 U.S. OPERATIONS, RESEARCH, AND DEVELOPMENT STRUCTURES

Monitoring all of Earth's environments for nuclear tests clearly requires a diversified effort that draws on scientific expertise in many fields. The U.S. capabilities for monitoring the CTBT are founded on a substantial infrastructure for data collection, analysis, and reporting that has been assembled over nearly 50 years of nuclear test monitoring. This includes extensive experience with seismic, hydroacoustic, infrasound, radionuclide (ground and airborne sensors), and satellite monitoring. The U.S. operational regime is supported by extensive basic and applied research activities that incorporates expertise from many agencies and the academic and private sectors. The monitoring efforts are further supported by the nuclear testing experience of the United States, as well as by a great variety of intelligence assets that serve to define monitoring capabilities in different regions of the world.

Although the full scope of U.S. national efforts is not described here, a few key roles are identified. Although nuclear test treaty monitoring is intrinsically an arms control issue, DoD and DOE have had primary responsibility for the U.S. effort for decades. Currently, the Air Force Technical Application Center has the primary operational task of monitoring nuclear testing treaties, and the U.S. NDC for CTBT monitoring is being established at AFTAC. AFTAC is implementing a computer analysis capability to access IDC and NTM data and carry out the detection, location, and event identification procedures that will lie at the heart of U.S. CTBT monitoring.

For more than 35 years, AFTAC operations have been strongly linked to DoD research programs, with basic research (the so-called 6.1 program), exploratory development research (6.2 program), and advanced development research (6.3 program) components. In recent years the 6.1 program has been directed by the Air Force Office of Scientific Research (AFOSR), the 6.2 program by the Air Force Phillips Laboratory (AFPL), and the 6.3 program by AFTAC. An innovative technology and advanced developmental program (6.2/6.3) has been directed by the Advanced Research Projects Agency (ARPA). These programs and their interrelations, are discussed at some length in the NRC (1995) report. Although some internal DoD research has been conducted by AFTAC, AFPL, and ARPA personnel, the primary research effort has involved peer-reviewed, externally funded university and private contractor researchers, with budgets of about $7.6 million per year provided by AFOSR, ARPA, and AFTAC in the past two years. In addition, ARPA has supported the development of a prototype-IDC, involving extensive hardware and software development, with a budget level of about $15 million year for the past two years. The latter effort involves advanced developmental research, which is separate from the fundamental research of the peer-reviewed program.

The Department of Energy has had primary responsibility for U.S. nuclear weapons development and testing programs and has also sustained long-term research programs that support nuclear test treaty monitoring. The latter include large internal research programs on seismological and hydroacoustic monitoring, satellite systems development, On-Site Inspection methodologies, and modeling of nuclear explosion signals in all media. In FY 1995, DOE was assigned an expanded responsibility for research and development for monitoring and/or verifying compliance with the CTBT, which encompassed all anticipated monitoring technologies and systems. This expanded DOE program is strongly linked to the AFTAC operational effort. For the past two years it has included a substantial ($3.665 million [FY 1995] and $4.395 million [FY 1996]) external funding effort supporting university and private contractor research activities. It appears that this external program will be greatly reduced beginning in FY 1997, leaving the DoD program as the main support base for university and industry fundamental research in support of the CTBT.

Many other government organizations contribute directly to U.S. nuclear test monitoring efforts. These include (1) the National Oceanic and Atmospheric Administration (NOAA), which has responsibilities for weather forecasting and atmospheric modeling, (2) the USGS, which receives seismic data and bulletin information from thousands of stations around the world and plays the lead role in documenting the seismicity of the United States; (3) the Office of Naval Research (ONR), which supports research in long-range acoustic propagation in the oceans; and (4) the National Science Foundation (NSF), which supports basic research in many relevant areas.[16] These other federal agencies support basic research in areas relevant to many of the CTBT monitoring technologies and can be viewed as indirect support of the U.S. monitoring capabilities. Earthquake monitoring conducted by the USGS-NEIS, the ISC, and many regional networks around the world provides independent determinations of event bulletins derived from much larger numbers of stations than the IMS. (2600 seismic stations currently report to the USGS.) These are a significant source of information about earthquake activity at large and small magnitudes that can support U.S. CTBT monitoring activities. As detailed in NRC (1995), this is an important example of the potential contributions to CTBT monitoring from research outside of DoD and DOE programs.

1.7 TRANSITIONS IN THE U.S. RESEARCH PROGRAM

In 1996, the DoD program was substantially restructured in response to changing priorities. Although AFTAC continues to be tasked with

[16] NSF also funds the Incorporated Research Institutions for Seismology (IRIS), which has deployed state-of-the-art global seismic stations through the USGS and the University of California, San Diego. Fifty of these stations are included in the Auxiliary Network and two are in the Primary Network of the IMS.

serving as the CTBT NDC, the AFOSR, AFPL, and ARPA research budgets were consolidated into a single DoD funding line, organized under the new Nuclear Treaty Program Office (NTPO), overseen by the Assistant to the Secretary of Defense for Nuclear, Chemical, and Biological Programs. The Department of Defense Appropriations Act of 1997 recommended a budget of $29.1 million for the associated Air Force arms control funding element, with $8.8 million for a peer-reviewed external CTBT monitoring research program ($7.1 million specifically for peer-reviewed basic research in the field of explosion seismology and $1.7 million for research in complementary disciplines such as hydroacoustics, infrasound, and radionuclide analyses). The other $20.3 million is for sustained development of the IMS IDC and involves operational systems development. The NTPO has tasked the Defense Special Weapons Agency (DSWA; formerly the Defense Nuclear Agency) to oversee the external research program, and an initial Program Research Development Announcement (PRDA) was issued by DSWA in November 1996. Initial contract awards are anticipated by mid-1997.

Consolidation of the DoD CTBT research and development program constitutes a major restructuring of the research community's support and has prompted widespread concern about future support for basic research in the field of seismology. There have been significant turmoil in the seismological research program over the past 15 years and philosophical disagreements over the balance and nature of the research program that will best service the CTBT monitoring effort. By eliminating past bureaucratic structures, DoD has an opportunity and a responsibility to set in place an effective CTBT research program for the future. Attendant issues are addressed in Chapter 4 of this report.

2

CTBT Monitoring Technical Challenges that Drive Research

INTRODUCTION

The extensive U.S. experience of testing nuclear weapons and monitoring foreign tests provides an essential foundation for identifying and locating explosions in all environments of the Earth. From first principles and testing experience, the explosion characteristics of nuclear devices are well understood. However, the interactions of these explosions with surrounding media underground, underwater, or in the atmosphere are less well known. Importantly, these phenomena also play a role in monitoring activities. Specifically, the interactions excite disturbances of sound waves, stress waves, light flashes, radioactive gases, and debris clouds that can be observed at large distances using sensors in the atmosphere, ocean, or ground. These disturbances propagate through the Earth system obeying well-known laws of physics. By continuously monitoring the diverse media on the Earth, one can detect arrival times and characteristics of explosion signals at different positions on the surface or in space. By using a fundamental understanding of the propagation processes, the signals at various stations can then be analyzed to locate and identify an explosion source. Thus, if the acoustic properties of the ocean were well known as a function of depth, temperature, salinity, and bathymetric structure, one could use three different observations of hydroacoustic arrivals to triangulate accurately on the source. In practice, however, such capabilities are limited because the properties of the Earth, oceans, and atmosphere are complex and not completely characterized.

Given these limitations, there are two principal challenges for CTBT monitoring. First there is a need for a sufficient distribution of global monitoring stations to ensure a high probability of detecting, locating, and identifying explosions at yields and confidence levels consistent with national goals. This is the primary challenge for designing, operating, and maintaining the IMS and the assets of NTM. Second, there is a need to analyze monitoring signals, using a complete understanding of transmission paths and phenomena, to allow location and identification of potential explosions with confidence amidst the significant background of natural and human-induced sources of noise (e.g., earthquakes, quarry blasts, undersea volcanic eruptions). This is the primary challenge for basic research in support of CTBT monitoring, and it is the focus of much of this report. For that reason, the basic monitoring procedures that are the drivers for the most important research problems are outlined in this chapter. Chapter 3 discusses the research activities necessary to enhance U.S. CTBT monitoring capabilities.

2.1 PHYSICAL PHENOMENA: SOURCE EXCITATION, SIGNAL PROPAGATION, AND RECORDING

Nuclear explosions can occur in space, the atmosphere, underwater, and underground. For monitoring sources in each medium the basic problem can be posed in terms of source excitation of the signals of interest, propagation of those signals through the various media, recording the signals, and ensuing signal analysis to detect, locate, and identify the source.

Atmospheric or Space Explosions

There have been 514 nuclear tests in the atmosphere and space conducted by five different nations (Adushkin, 1996). In an atmospheric nuclear explosion, a fireball is produced in the first fraction of a second after detonation as a result of the interaction of the atmosphere and the initial x-ray emission from the device. The thermal energy is reemitted in the visual and infrared spectrum, with the fireball light history involving a double flash (a rapid-duration flash followed by a longer-duration flash) that forms the basis for distinguishing nuclear explosions from chemical explosions and lightning. The pattern of light can be detected by satellite optical sensors (so-called bhangmeters) and is highly diagnostic of the source type. The atmosphere absorbs most other types of radiation (e.g. x rays, gamma rays, neutrons, beta particles) from low-altitude or below-surface tests, preventing them from being observed by remote sensors (DOE, 1993).

High-altitude explosions between 15 and 80 km produce large fireballs due to the thin atmosphere, with high visible light intensity. Above 80 km, the fireball is largely the result of ionized debris (DOE, 1993). As the atmosphere thins in approaching deep space (altitude > 110 km), the optical effects of the fireball diminish, but there is a corresponding increase in x-ray and gamma-ray emissions that can be monitored by upward-looking sensors on orbiting satellites. Explosions in the atmosphere produce ionized plasma in the source region that has a strong electromagnetic pulse (EMP) capable of disrupting communications over a large area and being monitored by EMP detectors on satellite systems. The DOE is currently developing a new generation of EMP detector for deployment as part of the U.S. NTM for atmospheric monitoring. However, numerous large lightning bolts as well as human-generated broadcasting activities also produce EMP signals that must be differentiated from explosion signals.

A nuclear explosion in the atmosphere will also produce acoustic signals (sound waves) that travel through the air at about 300 m/s, with energy concentrated in the 20 to 500 mHz range. Only a tiny fraction of the explosion energy is transmitted as sound waves, but testing experience has empirically been able to characterize the sound level as a function of explosion yield. This relationship, together with the background noise, provides a basis for defining infrasonic detection thresholds. Sounds transmitted through the atmosphere have predictable propagation properties influenced mainly by the vertical thermal and density structure of the atmosphere, which is quite well known. Secondary effects of wind patterns on the sound intensity are predictable if the wind patterns at various altitudes are known. IMS infrasound stations are arrays of surface atmospheric pressure sensors designed to record these signals.

In this period range, several natural phenomena can produce a background noise level. These include long-duration signals with a frequency of about 200 mHz, which are ocean wave-generated "microbaroms." Short-duration impulsive signals more likely to be confused with explosion signals are generated by volcanic blasts, meteor falls, sonic booms, and auroral infrasonic waves propagating beneath supersonic auroral electrojet arcs (Figure 2.1).[17] Historical experience with atmospheric monitoring suggests that at most a few events per day will be recorded at more than one of the IMS infrasound stations, whereas a 1-kiloton (kt) explosion will typically be observed at a large number of stations. Relative to the seismic problem, for which on the order of 100 events per day (or even more in the case of large earthquake aftershock sequences) are large enough for the IMS

[17] Auroral infrasonic waves are only recorded at high-latitude stations.

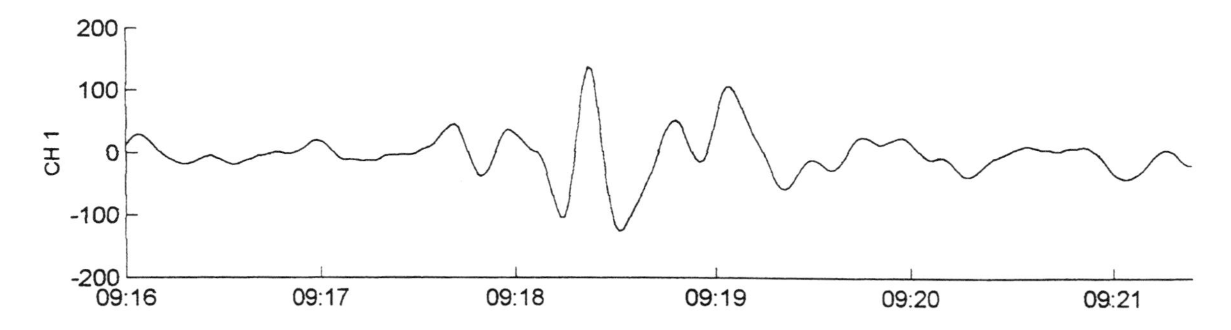

FIGURE 2.1 Example of a short-duration impulsive infrasound signal produced by an auroal infrasonic wave. The time scale of the record is approximately 5 minutes (Source: C. Wilson, personal communication, 1997).

to locate, the infrasound system will be confronted with far fewer events, but it could play an important role in identifying surface sources such as quarry blasts, as well as monitoring atmospheric events. The air disturbance from an atmospheric blast can also perturb the ionosphere, producing a propagating, large-amplitude deflection of the ionosphere that could be detectable by radiowave sensors, which reflect signals from the ionosphere, or by Global Positioning System (GPS) stations at the surface, which detect changes in the total electronic count associated with a frequency-dependent time delay. Such sensors are not part of the IMS.

Atmospheric blasts release radioactive particulates and gases that are transported by the wind. If the blast is near the Earth's surface, a cloud of dust, hot gas, and debris can rise with the vortex formed by the fireball, injecting debris into the upper atmosphere. Fission products and other vapors condense on soil and debris particulates, with the heavier particulates settling out near the source while lighter ones can be carried downwind (see Appendix G for more details). Explosions over water close enough to the surface for water to be vaporized by the fireball will have a large amount of local fallout as the fission products are washed out of the air, first with the descending water column, then with the resulting mist, and then in rain as the vaporized water recondenses (Glasstone, 1957). Radionuclide detectors at the surface will detect the particulates and radiogenic noble gases that are produced by atmospheric nuclear explosions, with wind patterns determining which stations "see" the event and at what time. Wind transport velocities are relatively slow (compared to sound waves); thus, radionuclide detections may take days to weeks, depending on source location, weather patterns, and the location of the nearest fixed station. The radioactive decay with time and the reduction in concentration due to various mechanisms as the material spreads out reduce the detectability of the material. Rain-out can rapidly reduce particulate concentrations, so the opportunity to observe the debris is time-limited. Rain-out will have little effect on the concentration of radioactive noble gases; so the monitoring of these gases is important. Radionuclide observations can be reliable indicators of an atmospheric test if the material is collected and analyzed early enough.

Near-surface atmospheric explosions can couple energy from atmospheric sound waves into the ocean or solid Earth. Thus, there is some role for hydroacoustic and seismic monitoring of explosions in the atmosphere, particularly in the case of evasion scenarios that attempt to mask the explosion by near-surface detonation (e.g., testing near the surface on a cloudy day with strong rains and certain masking efforts).

Underwater Explosions

Underwater nuclear tests are a major concern because, like atmospheric tests, they could potentially be carried out in remote international waters, where—even if detected and identified—they may defy attribution (see Section 2.2). There have been at least eight underwater nuclear tests (five by the United States, three by the U.S.S.R.), which provide some basis for understanding these events. For a device tested underwater at depths less than about 70 m, the fireball will be significantly reduced relative to an atmospheric test, but most of the radioactive noble gases and a significant fraction of the other fission products will be released into the atmosphere, with the remainder deposited near the water surface. The altitudes reached by the ejected particulates tend to be lower than for low-atmosphere shots, and radioactive fallout tends to be reduced relative to explosions above land due to the lack of larger particulates. Tests in the depth range 70 to 300 m will allow a significant fraction of the noble gases to be released, but most of the other fission fragments will remain in the water. If the test is conducted deeper than 300 m, it is possible that no fireball will be detectable and most of the radioactive noble gases, along with the other fission fragments, will be absorbed in the ocean water before the bubbles can reach the surface. For such tests, there will be efficient excitation of acoustic and seismic signals. In underwater tests, much of the radioactive material will accumulate in the thermocline, a thermally stable layer of water that exists in the oceans from depths of about 200 to 900 m. A seasonal thermocline at a much shallower depth forms during summer as a result of solar heating. Thus, a highly radioactive pool will be formed that disperses much more slowly than do atmospheric clouds of radioactive gases and particulates.

An important effect of the hydrostatic pressure in the water column is that hot gases released by the explosion are contained in a bubble, which typically expands and contracts in radius several times and rises before collapsing. The repeated oscillations give secondary pressure pulses that generate acoustic waves in the water. These are called "bubble pulses" (see Appendix E for further discussion). Near-surface events may lack a bubble pulse because the hot gases vent directly to the surface.

Monitoring underwater explosions involves a combination of satellite, infrasound, radionuclide, seismic, and hydroacoustic methods, with the latter two playing the largest role. The explosion and repeated pulses of gas bubble oscillation produce sound waves that spread into the surrounding ocean. The variation of sound velocity in the water is sensitive to temperature and pressure conditions. A low-velocity waveguide, the SOFAR channel exists in almost all ocean environments to varying degrees (in some regions the minimum sound velocities are right at the ocean surface, but typically they are 0.5-1.5 km deep). Sound is channeled into the SOFAR channel for sources within, above, and below the waveguide, and sound waves with frequencies below 200 Hz can travel thousands of kilometers in all directions with little attenuation, although variations in seafloor topography, water temperature, and salinity, along with islands and seamounts, can block or modify propagation along some paths. The specific excitation of sound waves depends on the explosion depth, the local structure of the sound velocity profile, and lateral gradients in the velocity profile. Hydrophones, typically deployed vertically in arrays across the SOFAR channel or in some cases deployed horizontally in arrays on the ocean bottom, can detect the passage of hydroacoustic waves and measure the corresponding pressures in the sound pulses and their propagation direction (if arrays are used, but not for single sensors).

A 1-kilogram explosion detonated in the center of the SOFAR channel in most regions of the open oceans will be detected readily at numerous worldwide hydrophone systems. This results from the remarkable efficiency of sound propagation in the oceans. Explosions at different positions with respect to the SOFAR channel will couple less efficiently, but most excitation effects are well understood except for sources right at the ocean surface or ocean bottom. At these interfaces the synergy between hydroacoustic methods and infrasound, seismic, and NTM is an important factor. In principle, it is possible to detect events anywhere in the ocean down to fractions of a kiloton in yield, as long as the hydroacoustic network is adequately configured. A sparse hydroacoustic network intrinsically has blind spots that must be monitored by other means, due to blockage of hydroacoustic signals by coastlines or ocean bathymetry. If the network is composed of single sensors that cannot measure propagation direction, the association of arrivals from a given event in the presence of nonstationary noise presents special problems.

Aside from blockage of the SOFAR channel for

certain source regions, the main operational issue for a hydroacoustic network is discrimination of the various noise sources from explosions to avoid false alarms and to provide the location of sources of all types. Given the efficient sound transmission properties, low-amplitude acoustic signals are detectable around the world from chemical explosions for military exercises and airguns for oil exploration. Underwater earthquakes can also produce vibrations of the ocean floor that couple into hydroacoustic signals (T-phases). These tend to be long-duration signals due to the size of the effective source region where they are excited (the surface area over which seismic wave energy couples into hydroacoustic energy). Underwater volcanic explosions can emit impulsive sound waves, either in short bursts or in long sequences (sometimes called "popcorn" signals), and bubble pulses may be produced by some volcanic eruptions if magma is exposed to ocean water. Meteorite falls can produce significant hydroacoustic signals if they impact the water. Ships produce both narrowband and broadband signals, and whales produce noises at low frequencies as well.

Given the effectiveness of low-level detections in the hydroacoustic environment, a major operational decision involves determination of the sound amplitude threshold below which events will not be considered in order to avoid being swamped with analysis of numerous small sources. This requires an informed assessment of the extent to which source excitation can be reduced by testing at the surface or in regions with extensive blockage of the SOFAR channel, as well as the synergies achievable through the use of other monitoring technologies, including NTM.

Seismic stations contribute to monitoring the underwater environment in several ways. First, explosions in the water are efficient generators of seismic waves, because hydroacoustic energy efficiently generates downgoing seismic waves below the water. As a result, seismology provides monitoring for regions in which a hydroacoustic network has blind spots. Second, submarine earthquakes that generate T-phases should also generate seismic waves whose approximate origin time and location can be used to identify the hydroacoustic signal (which in turn may help to locate the event). In addition, seismic stations can be deployed on the ocean bottom or in drill holes in the ocean floor, and these can directly observe both seismic and hydroacoustic signals (changes in water pressure due to sound waves in the ocean move the ocean bottom up and down and thus are recorded on the seismic ground motion sensors). Seismic stations deployed on islands and near continental margins can detect hydroacoustic waves that convert to seismic waves at the land margins and then propagate through the rock. The efficiency of conversion of hydroacoustic to seismic energy laterally across a coastline is strongly influenced by the slope of the underwater margin, with steeper slopes favoring coupling. Five of the IMS hydroacoustic network stations are seismic stations deployed on islands for the purpose of monitoring hydroacoustic phases converted to seismic T-phases. Typically stations must be deployed on several sides of an island to provide sensitivity to waves arriving from different directions.

It is important to recognize that as a result of political considerations related to concerns about tracking ships and submarines, the IMS hydroacoustic network was designed to be a sparse, limited-capability system. For example, IMS hydroacoustic stations involve only single sensors, rather than distributed arrays. The resulting inability to determine the direction of approach of hydroacoustic waves leads to difficulty in associating signals with particular sources and in determining source locations. Furthermore, the sparse distribution of IMS hydroacoustic systems results in many blind spots due to blockage of propagation in the SOFAR channel by coastal configurations and variations in bathymetry. These blind spots will require seismic monitoring or coverage by NTM.

Underground Explosions

Underground nuclear explosions present a monitoring challenge. Although there has been extensive underground testing, with more than 1570 known tests (Adushkin, 1996), and the physics of excitation of seismic signals by explosions is quite well understood, the solid Earth is physically a heterogeneous and "noisy" environment. Because the solid Earth is much less accessible for direct measurement of physical parameters than the oceans, atmosphere, and space, models for seismic propagation in the interior are correspondingly less detailed. Furthermore, numerous non-nuclear sources produce seismic signals similar to those expected from small nuclear explosions, so that the potential for event mis-identification (false alarms) is substantial. These factors pose the greatest research challenges to seismology; however, hydroacoustics, infrasonics, radionuclides, and satellite imaging all can contribute to monitoring underground tests; thus, research in these areas is also warranted. The OSI issues for the underground environment are also challenging.

Underground nuclear tests are typically designed to be well contained, meaning that the depth of burial, the material properties, and other operational factors are sufficient that explosion materials do not vent to the surface. However, many underground tests in the past did in fact vent radionuclides. Presumably, a determined evader would want to avoid significant release of debris in order to avoid detection by radionuclide monitoring or OSI. The evader would also want long-term containment of noble gases, which is difficult because of rock fracturing above the explosion site and the mobility of these gases.

For a 1 kt explosion, U.S. experience suggests that burial depths of greater than 107 m will contain gases from the experiment; the depth of burial for containment scales with the cube root of the explosion yield (DOE, 1993). Underground explosions at these depths or greater may be located in boreholes, tunnels, mines, or caves. If drill holes are used, activity related to drilling, emplacement, and diagnostic facilities may be detected and monitored by high-resolution overhead methods, although such approaches typically work best when the location of the test is well known. Emplacement in mines or tunnels or in large cavities excavated in rock or salt for the purpose of decoupling could allow concealment of diagnostic facilities. Unrelated mining activities might further obscure nuclear testing operations and signals.

An underground nuclear explosion releases energy in microseconds, with a fully tamped explosion vaporizing rock surrounding the device and producing a gas-filled cavity having pressures of several million atmospheres that expands outward rapidly, sending a compressional shock wave into the surrounding rock. This shock wave initially melts and shatters the medium as it propagates but progressively decreases in amplitude until the rock begins to respond elastically (i.e., it returns to its initial state after the disturbance passes by) to the outward-propagating wavefront. At this point, called the elastic radius, the wave becomes a type of seismic wave called a P wave (analogous to a sound wave in the air) that propagates away from the source volume as a spherically expanding wavefront of radial compressional motions of the solid rock. The explosion source process is rapid relative to natural earthquake events of comparable total energy release, and in the ideal case for a homogeneous source region the seismic radiation from an explosion is isotropic, with uniform amplitudes over the spherical P wavefront. The physical dimension of the source volume (defined as the region of nonlinear rock deformations within the elastic radius) is much smaller than the fault zone for a natural earthquake of comparable energy release; however, this difference decreases at small magnitudes, and attenuation may eliminate the frequencies at which the differences could be observed. For both explosions and earthquakes, only a small fraction of the total energy released at the source is contained in the elastic wavefield.

Solid materials transmit a second type of elastic wave that involves transient shear deformation of the medium. Known as S waves, these propagate only through the solid Earth, not in fluids such as the oceans or atmosphere. Both P and S waves are referred to as body waves because they travel through the Earth's interior. Although P waves are the predominant seismic signal from explosions, a smaller fraction of S waves can be generated from asymmetries in material properties around the source, cracking and shearing of rock above the shotpoint, and relaxation of ambient tectonic stresses in the rock volume surrounding the explosion cavity. The S waves travel more slowly than P waves: for a typical crustal rock the P-wave velocity is on the order of 6 km/s, whereas the S-wave velocity is 3.5 km/s.

Pressure and temperature effects, along with changes in rock mineralogy and composition, cause seismic velocities to vary with depth as well as laterally at a given depth. At the Earth's surface and at sharp internal boundaries between rock masses with different seismic velocities and densities, P-wave energy can convert to S-wave energy and vice versa, so the total elastic wavefield in the Earth always involves a complex distribution of propagating P- and S-wave energy, with the source and propagation effects controlling the partitioning of energy. At the surface, the interaction of P- and S-wave energy produces disturbances that travel as surface waves (i.e., their amplitudes decrease with depth from the surface). Seismic wave energy is also dissipated by irreversible processes such as friction, fluid pumping, and dislocation motions in crystals. These mechanisms attenuate seismic waves with increasing propagation distance and spread the seismic energy over larger and larger areas of the Earth's surface.

Seismic waves produce characteristic motions of the medium that can be recorded by ground motion sensors (seismographs) located at fixed positions. By recording three orthogonal components of ground motion as a function of time, the complete history of ground motion at the seismic station can be recovered for analysis, albeit limited by the filtering effects of the instrument. The times of particular ground motions are controlled by the distance from the source, the time of the source event, and the seismic wave velocities between the source and the receiver. The recorded ground motions, or seismograms, are time series that are typically digitized with discrete time sampling and written to tape or disk or telemetered to a central analysis facility. A station at a given distance from a source will record a sequence of arrivals involving P and S waves that travel through and reverberate in the Earth's structure and surface waves that travel along the surface from the source to the station.

The general characteristics of seismic waves differ as a function of the distance traveled. Signals recorded at distances less than 1000-2000 km are called regional waves, whereas those at distances greater than 2000 km are called teleseismic waves. Regional waves primarily travel in the low-velocity crust (25 to 70 km thick in continental regions) and uppermost mantle. Seismic wave amplitudes at regional distances are strong near the source but decrease in amplitude with distance and can be too small to detect beyond 1000 km. Thus, the word "regional" here carries the additional implication that such waves are dependent on the properties of the Earth's crust and uppermost mantle, which can vary quite strongly from one region to another. The P waves in the distance range 1000 to 2000 km tend to have low amplitudes due to defocusing by upper-mantle velocity gradients and/or waveform complexities due to interactions with the velocity layering in the transition zone (400-670 km deep); this range is sometimes called the upper mantle triplication range. Teleseismic body waves tend to penetrate through the mantle transition zone and reach their lowest point in the lower mantle or core of the Earth. Teleseismic P waves from 3300 to 9000 km tend to exhibit simple propagation effects and are the most straightforward to interpret.

For decades following the LTBT, when nuclear testing was carried out underground and in-country monitoring was not permitted, monitoring was conducted using teleseismic signals. With the CTBT and the permitted use of in-country stations, attention returns to the study of regional waves because they provide the strongest signals and may be the only detectable seismic signals from small-magnitude sources. Though regional waves can have large amplitudes (and thus be detected easily if a seismometer is operated at a regional distance from a source of interest), they are more complex and harder to interpret than teleseismic waves. An extensive data set of seismic signals from a given region must first be acquired and understood before regional waves recorded within it can be interpreted in detail, a process frequently referred to as calibration. (Sometimes, however, the data from a more easily accessed region can be used to aid in the interpretation if the two regions are geologically similar.) The IMS will rely mainly on teleseismic data to achieve global monitoring, whereas NTM will have to use regional data for certain areas in order to meet U.S. monitoring objectives. Thousands of stations outside the IMS are potentially available to supply supplementary regional data for events that the IMS and NTM may detect and to aid in the calibration of IMS and NTM stations.

A major seismic monitoring challenge arises from the fact that ambient ground motion forms a background noise level at every seismic station. This noise is composed of surface motions caused by local weather conditions, vibrations produced by wave interactions in the oceans (microseisms), human-induced activity, and many other localized or distributed sources of seismic waves. It is station specific and establishes a lower bound for detectability of P and S waves from distant sources. Recording signals with groups of sensors distributed in a pattern—a seismic array—allows the signals to be enhanced by summing the traces, thereby suppressing incoherent background noise. Array recordings can be processed to determine the direction of approach of a wave as well.

A great number of background signals are as large as the signals expected for a small nuclear test. For example, there are about 7000 earthquakes per year with seismic magnitudes of 4.0 or larger and about 68,000 earthquakes per year with magnitudes of 3.0 or larger (Ringdal, 1985). A global seismic monitoring system that compiles all global events larger than magnitude 3.0 would have to detect, locate, and identify 186 events per day, on average, as well as with deal with intermingled arrivals from many smaller events detectable at some stations. Pushing the monitoring threshold downward has serious operational implications because there are about 209,000 magnitude 2.5 earthquakes per year (572 per day). At the lower magnitude, quarry blasts and natural phenomena such as volcanic eruptions, avalanches, and landslides add to the total number of events that can be detected. Fortunately, U.S. monitoring concerns greatly reduce the areas that have to be monitored down to low seismic thresholds, which greatly decreases the number of events that must be processed in the CTBT monitoring system.

The coupling of seismic waves into acoustic waves in the oceans and atmosphere, as well as the direct excitation of infrasonic waves by sources that disrupt the surface, allows hydroacoustic and infrasonic methods to be used to monitor underground events. Other methods that exploit energy coupling from the solid Earth to the atmosphere could also be useful. These include monitoring of ionospheric deflections, EMP, and radioactive materials released by venting. In addition, postevent seepage through cracks and fissures might be detectable by OSI methods (which can include drilling back into the source region). The upgoing shock wave can affect surface flora and soil conditions, displace material at the surface, and cause faults and fissures to appear. If the surface material settles back into the explosion cavity, a collapse crater may appear at the surface. Such surface disruptions can be detected by satellite imagery and OSI methods. Highly precise pre- and postexplosion imaging is needed, so this is a practical monitoring strategy only for localized candidate testing areas that are predefined by other criteria. Finally, any underground nuclear test will involve significant drilling, tunneling, mining, and instrumental recording activities that may be monitored by satellite imagery and other NTM, again subject to the need to localize the activity by methods other than imagery.

2.2 FUNCTIONS OF MONITORING SYSTEMS

Section 2.1 summarizes the physical processes involved in exciting wave disturbances and radioactive emissions that potentially can be detected at large distances. The process of monitoring the CTBT involves systematic screening of all types of signals detected by the monitoring network for the purpose of identifying those from any nuclear explosion that might take place. The challenge lies in distinguishing signals generated by natural or benign human-induced sources from those of a nuclear explosion against a background of ambient noise. At every stage of the monitoring operation there are basic functions that must be performed that depend on scientific understanding of the phenomena; these are discussed below.

Event Detection

The first substantial signal processing step in CTBT monitoring by technical means is the detection of signals of the type that would be emitted by a nuclear explosion. In the case of seismic, hydroacoustic, and infrasound technologies, this means searching a continuous time series for excursions from or changes in the background "noise" level or the appearance of spatially correlated energy with little or no change in signal level or spectra. In the case of radionuclide technologies, it means comparing the gamma-ray spectra collected from daily samples with the spectra expected from products of a nuclear explosion. The remainder of this section discusses the processing of the continuous time series. The seismic methodology forms the basis for this discussion because it is most developed in the CTBT context. Processing is similar for the hydroacoustic and infrasound technologies.

Each underground event creates a variety of signals due to the near-source effects described in the previous section and the heterogeneity of the paths from the source to the receiving site. The pattern of signals varies from station to station, and the signals from the many nonnuclear sources overlap and mask each other. The first step in signal processing is to determine if individual signals are present and then to group together those signals associated with a specific event. In the case of both hydroacoustic and infrasonic systems, the propagation speed is lower and the presence of spurious signals can lead to difficulties in associating signals, particularly for sparse networks.

The seismic signal detection process at the prototype IDC combines signal tuning, a short-term/long-term detection algorithm, and processing for array stations to detect the waves and determine the arrival times to be used in the association algorithms. Because of the large volumes of data, automatic detection processing is essential. The signal detection processing relevant to CTBT monitoring must address the problem of detecting unknown or highly varying signal shapes with spectra that overlap the spectra of the background noise. Publications of the Group of Scientific Experts of the Conference on Disarmament and of the Center for Monitoring Research describe the process and its results in detail. The following discussion is an abridged version of those descriptions.

Signal Tuning

Tuning begins with the identification and adjustment of bad data because failure to do so results in many spurious detections and, ultimately, failed or false associations. The individual channels are examined in 4-second segments for spikes and for sequences of zeros or data of constant value. If the number of data with bad values in the segment does not exceed a number specified by the user, the bad data are adjusted by setting them to zero or by interpolation between adjacent values. If the number of bad values in the segment exceeds the reference number, the segment is excluded from further processing.

Detection Algorithms

The data used in the detection process consist of filtered versions of tuned individual or, in the case of arrays, combined channels. The criteria for combining the data reflect the information known about the phase velocities, frequency content, azimuthal distribution, and optimum channel weights at each station. Coherent and incoherent array beams (beams are sums of seismograms with small timing shifts to align them on a particular arrival azimuth and angle of incidence at the surface; coherent beams sum the actual seismograms, whereas incoherent beams sum envelopes of the seismograms, removing the phase information) are used to detect the P and S waves, respectively. The IMS beams are equispaced in azimuth and often are based on subsets of array channels in order to suppress coherent noise. Various steps in the process are applied to the data from both array stations and three-component stations.

In the CTBT context, seismic signal detection processes are currently tuned to detect the first arrival of a series of seismic phases associated with an event. The tools are applied in the frequency range 0.5-5 Hz. The most common approach is to compare some measure of the short-term average (STA) signal amplitude or power to a similar measure of the long-term average (LTA) signal amplitude or power. When the STA exceeds the LTA by a fixed level, a detection is declared. It is also common to make this comparison in several rather narrow-frequency bands, a detection being declarable on one or on a weighted combination of filtered channels. In arrays, a detection will often be declared on several beams nearly simultaneously. The beam with the highest signal-to-noise ratio is accepted as the correct one. Signal detection algorithms used in some networks require detection of the data from more than one element of the network before a "network detection" is declared. This is an effective criterion for eliminating false or spurious detections, but its use raises the detection threshold because detection must occur at the station with the worst detection level of the stations used.

Detection parameters must be adjusted for each array or three-component station to achieve optimum performance. The primary adjustable parameters are the durations of the STA and LTA, the STA/LTA ratio above which a detection is declared, and the filter bands. For example, studies of the station at Lormes, France, indicated that a reduction in the length of the interval used in computation of the short-term average reduced the number of false alarms while detecting more actual seismic phases. In general, these parameters are adjusted so that few, if any, signals from real events are missed. The result of this approach is that many, up to 90 per cent, of detected signals are never used either because they are spurious or because they cannot be associated with other detections to form a well-located event. Nonetheless, experience with the prototype IDC in December 1995 showed that about 30 per cent of the detections used in event location must be added by the analysts, so there is significant room for improvement in detector efficiency.

Onset Estimation

Once a signal detection has been declared in an automatic processing system, the next step is to measure the characteristics of the detected signal. This includes amplitude, period, azimuth of approach, direction of ground motion (in the seismic case), phase velocity, and time of arrival or onset time. An accurate onset time is critical for the location process; the other characteristics are useful for source location and for identification of the arrival type and the source.

At the prototype IDC, determination of the onset time associated with a detection is usually made on the detecting beam with the largest signal-to-noise ratio. If the detection is made on an incoherent beam (which is usually the case for S-wave detections), the onset is estimated from a vertical coherent beam composed of a subset of filtered data streams similar to those of the detecting beam. The time of arrival of the onset is defined as the time corresponding to one-quarter cycle before the arrival of the first large peak in the window surrounding the initial detection. The results from processing have agreed with those of an analyst to within 1 second about 75 per cent of the time.

Other techniques have been proposed. The prototype IDC has an option to apply a "Z" transform based on a statistical treatment of the excursions of STA from the mean level. Another technique developed by Pisarenko et al. (1987) and Kushnir et al. (1990) defines the onset time as the time at which the statistical features of the power or frequency content of the observed time series change abruptly. This technique uses maximum likelihood spectral estimates of the waveform in an autoregressive model. The experience in the prototype IDC, using autoregressive models, was that most of the arrival times were late, and manual corrections moved the automatic arrival forward in time, up to 2 seconds. However, about one-half of the automatic time "picks" remained unchanged when reviewed by an analyst. More research is required to refine these techniques in the seismic case.

Other deterministic techniques, described by Stewart (1977), work entirely in the time domain. They transform the seismogram into various versions of a "characteristic function" or recombination of the individual data points in which abrupt changes in amplitude or frequency are more easily recognized.

Association

Whereas the detection of arrivals in a given time series requires relatively little knowledge of the medium in which the signals are propagating, the process of associating arrivals at different stations with a common source requires extensive knowledge of how signals traverse the medium. At first thought, this seems like a straightforward activity, but in fact it typically drives the most computer-intensive operational element of any monitoring system. The problem is complicated by the mixing of detections from many global sources, and some detections may be observable at only one or a few stations. The systematic variation of signal amplitudes with propagation distance must be allowed for because a strong arrival at a station may be produced by a small local event, a moderate-size event at a greater distance where wave energy gets strongly focused, or a distant strong event. In all media, but particularly in the solid Earth, multiple arrivals from a given event are expected in a distance-dependent sequence, and these may or may not be detected and identified completely at each station (due to variations in source radiation pattern, source excitation effects, attenuation and blockage, geometrical spreading effects, and detector sensitivity). Associating radionuclide anomalies with a source region can involve either backtracking wind patterns from observing stations or using forward calculations for possible event locations based on other monitoring technologies. Particular challenges for association occur when there are clusters of events, such as earthquake aftershock sequences and chemical explosions in some active mining areas, whose signals must be unscrambled to locate each event. In the oceans, seismic exploration for hydrocarbons and shipping can mask some events of interest to CTBT monitoring, such as low-altitude atmospheric explosions over water. Furthermore, the small number of hydroacoustic stations, the low speed at which acoustic waves propagate, and the multiplicity of transient sound sources in the ocean pose a special problem in association for ocean paths.

The standard information used to associate events is empirically determined travel times of different wavetypes as a function of distance from the source, or so-called travel time curves. Observed travel time curves are often used to determine models of the velocity variations in a medium. These models are then used to predict travel times for all phases (even those that are poorly observed empirically). If the medium is so heterogeneous that a single travel time curve cannot be used everywhere, regional travel time curves can be used (or a laterally varying Earth model can be determined) or empirical corrections for travel times in different parts of the Earth can be tabulated. Travel time curves for direct teleseismic P waves in the Earth typically predict global arrival times to on the order of ±1 second at a given distance (greater scatter is found at regional distances), so association is based on this level of consistency of arrivals at different stations for any postulated event location and origin time. Classic procedures for seismic association would use pairs of detections to form trial location estimates that were tested for consistency with other arrivals, both in the positive sense of having a detection with a consistent arrival time and in the negative sense of not having a detection when one should have been observed for the trial location.

A critical parameter in the association process is the definition of the number of detections required to form a legitimate event, especially given a sparse network intended to monitor low thresholds. Phase identification thus plays a major role because confident identification of multiple arrivals from a single event at a given station can reduce the number of stations needed to define and locate the event. Analysis of the direction of ground motion provides the primary basis for identifying seismic wave arrivals, which have distinctive angles of incidence depending on their propagation path and distinctive polarization of ground motion depending on the type of phase. Association is also influenced by the extent to which individual detections can be processed to characterize the azimuth of arrival at the receiver. In principle, an event can be defined and located by a single seismic station or array if at least two phases (e.g., P and S) are detected, associated, and identified correctly, providing an estimate of the distance to the source based on the systematic variation of relative arrival times as a function of distance and the azimuth to the source is determined by polarization or array analysis. In practice, multiple stations are often used because associating detections from multiple stations reduces errors in the location estimates. The prototype IDC currently uses an automatic event definition based on a sum of weighted observations at Primary stations, with a weighted sum threshold of 3.55 and no requirement for the minimum number of stations. All automatic event lists contain some events with one or two stations, but the final Reviewed Event Bulletin (REB) produced by the prototype IDC includes only events defined by at least three P-type phases at Primary stations.

The actual process of associating multiple detections at the prototype IDC involves a recent research development called Global Association (GA). A global grid of possible source locations is considered systematically as a function of time by testing the list of station detections for consistency with predicted arrival times at each station. This exhaustive search, enabled by fast computer processing, is more effective than classic methods of iteratively forming and breaking up trial location estimates. It has the advantage that a priori knowledge about the time corrections, blockage characteristics, and noise levels for various possible source-receiver geometries can be incorporated. This strategy can be implemented for all of the monitoring technologies if the a priori knowledge is available. Current versions use parametric measurements for discrete detections, but research is being done on methods that use the continuous time series. Algorithms proposed by Ringdal and Kvaerna (1989), Shearer (1994), and Young et al. (1996) operate on the waveforms themselves, without relying on the determination of times of arrival of individual phases. Such methods blur the distinction between detection and association and, in effect, detect events rather than individual phases. These methods show promise, but more work must be done before they can be introduced for general-purpose processing.

There is a strong interaction between detection thresholds and the association problem. During December 1995, only 5 per cent of the detections at IMS Primary stations and 6 per cent of the detections at Auxiliary stations were eventually associated with events that were published in the REB. Presumably this reflects large numbers of small arrivals detected at single stations and is a necessary outcome of monitoring low-magnitude thresholds. One cannot simply restrict the detections of interest to those with certain amplitude levels, given the variability of signal amplitudes caused by source and propagation effects.

The methodologies for associating hydroacoustic and infrasonic detections are similar to those for seismic signals. In the study of underwater acoustics, however, different terminology has been used, and there are some differences in signal processing details. For example, the process of association in acoustics would normally include array processing and interarray processing. Acousticians have typically dealt with arrays of individual sensors and relied heavily on coherent array processing and less heavily on coherent interarray processing. The IMS hydroacoustic network has no array processing involved in the conventional acoustic sense because there are no arrays. Thus, association of the signals at widely separated hydroacoustic and T-phase stations to triangulate on event location is identical to the seismic process.

Source Location

An event location is given by the latitude, longitude, depth or altitude, and time at which the energy release occurred. The association process generates an approximate location for an event, which must then be refined. It must be recognized that in the absence of ground truth for a given event, there are only estimates of the actual source location. Typically, this is obtained by solving a mathematical problem where the data are the arrival times of various types of signals at different sensors around the Earth's surface and the "model" is a set of relationships between travel times and distances for various wavetypes. In practice, the prototype IDC uses an iterative, nonlinear inversion process involving measurements of the arrival time, the azimuth of approach, and the apparent velocity of the wave along the surface.

Consider the situation for an event to be located by seismic waves. Estimating the source parameters most consistent with a group of associated arrivals requires source depth-dependent travel time curves for the Earth, which is one form of "model." This empirical information can be tabulated or incorporated into another form of "model" involving an explicit representation of the seismic wave velocity structure inside the Earth obtained from inversion of the travel time curves. Such inversions essentially smooth and interpolate the surface observations, giving a mathematical representation that can be used efficiently in computing theoretical travel times for waves anywhere in the medium (essentially giving the same result as interpolating travel time curves directly). Velocity models can be prescribed as a function of depth only (one-dimensional models) or as complete three-dimensional structures, while travel time curves may be regionalized or accompanied by source region or station corrections that account for deviations from a given curve. However detailed they may be, the resulting seismological models are simplified and averaged representations of actual seismic velocities inside the planet. Imperfect knowledge of the three-dimensional distribution of seismic velocities intrinsically limits the ability to determine the source's location and origin time, given a set of observed P and S arrival times at different stations (see Appendix C).

All underground source locations are estimates that depend explicitly on the seismic model used to interpret the observed arrival times, as well as on the extent to which available observations intrinsically allow the location to be estimated (e.g., the azimuthal distribution of stations recording the event). Uncertainties in source parameter estimates therefore involve the combined effects of using an approximation to the actual Earth structure, measurement error, and limitations due to station coverage. Estimates of uncertainty are usually given in terms of a "90 per cent error ellipse," defining that area of the Earth's surface within which the located event is expected to fall with 90 per cent confidence. There is a fundamental difference between the precision of a location estimate and its accuracy. Precision reflects the random variability in the solution, whereas accuracy pertains to errors relative to the true location. Internally consistent data sets can give exceedingly precise locations that are badly biased due to the use of an incorrect model; these therefore have low accuracy. The uncertainty due to measurement error and station or phase coverage can be assessed formally in the context of the mathematical inversion procedure, but the biasing effects due to inaccurate propagation models are much harder to determine. For small events that are recorded by only a few stations, typically at regional distances where the waves travel in heterogeneous crustal structure, uncertainties in the velocity structure usually result in poor location estimates, and there are corresponding difficulties in identifying the source type. In 1995, 18 per cent of the events reported in the REB had error ellipses with areas less than 1000 km^2 (the CTBT goal); 49 per cent had error ellipses between 1000 and 10,000 km^2; 23 per cent had error ellipses between 10,000 and 100,000 km^2; and 10 per cent had error ellipses greater than 100,000 km^2 (Ad Hoc Group of Scientific Experts, 1996). Deployment of additional IMS stations will improve performance significantly, particularly as azimuthal coverage improves and station corrections are determined, but the error ellipse estimates may not reflect systematic or model uncertainties.

Statistical estimates of the model uncertainty can be extracted from large data sets or from physical bounds on viable model parameters, but these approaches tend to result in unacceptably large location uncertainties. Typically some form of direct calibration is required, involving, for example, travel time corrections for paths from a source region to the monitoring network that account for errors in the standard model used. Calibration can be performed by using events with known source parameters (e.g., controlled explosions or earthquakes that rupture the surface) or, less directly, by developing improved three-dimensional velocity models that reduce model inadequacy (Appendix C considers this issue in some detail). A major problem confronting CTBT monitoring is that uncertainties in current seismic velocity models (even three-dimensional versions) are large in the context of the accuracy desired, and the opportunities for direct calibration of source or receiver paths are limited. This is true for hydroacoustic and infrasound methodologies as well, but the solid Earth environment is significantly more inaccessible in terms of measuring the properties required to make a highly accurate model of the medium. Thus, achieving the CTBT requirement of high-confidence absolute location uncertainties of 1000 km^2 requires both accurate seismic models or network calibration and a clear understanding of the remaining biases in absolute location accuracy, along with use of complementary technologies such as satellite imaging. Although the acoustic velocities in the infrasound and hydroacoustic methods can, in principle, be measured directly, these velocities are time dependent, with major dependencies on seasons and—in the case of the atmosphere—on winds. Radionuclide detections can contribute to source location by providing information about the strength and isotopic distribution at various sites. Atmospheric back-trajectories can be computed to provide low-resolution estimates of the source location that can then be correlated with events detected by other methodologies. This procedure would take place days to weeks after the event was located by seismic or acoustic signals.

The event's location estimate includes a value for its depth, which can be the most poorly determined parameter for most events. It reflects the fact that variations in depth of the source have similar travel time effects at all stations and thus there is a strong trade-off between the depth of the event and the origin time. The prototype IDC assumes an initial location at the surface and changes this only if the data absolutely require it. Seismic depth phases (waves that bounce off the surface above the source and hence arrive at the stations with delay times that indicate depth) can resolve the depth, but during December 1995, less than 10 per cent of the REB events had depths determined by depth phases (which are clearest for relatively infrequent earthquakes deeper than 50 km). During the same time, 25 per cent of the REB events had depths resolved by the location process (Ad Hoc Group of Scientific Experts, 1996). Calibration of the network will improve depth resolution, which is desirable because it is one of the definitive ways of ruling events out as possible nuclear tests.

The principal method for improving seismic event locations by the IMS and U.S. NTM is will be a long-term process of comparing the locations with those from various regional, national, or international organizations operating large numbers of stations. The panel notes that this process would be enhanced if IMS data were openly available to the research community. The underlying point here is that for treaty monitoring, location estimates must be provided promptly, within one or two days, which typically prevents making use of numerous additional stations that have good data but do not report to the IMS or the U.S. NDC. Fortunately, many event locations with precision better than 5 km are published openly (although not always promptly) for many regions of the world, and these locations can be used to build up an archive of calibration events with which treaty monitoring organizations can improve their location procedures.

Given multiple monitoring technologies for which the data streams can be fused, event location can incorporate the propagation attributes of multiple wavetypes if detections are associated reliably. This presents some challenging research problems for merging different data types with different intrinsic resolution. With radionuclide data being delayed relative to other data types, source regions determined by backtracking can be used to identify sources detected by other means or as a motivation to reexamine archived data for undetected events.

Size Estimation

Interpreting the strength of signals recorded at large distances from a source in terms of the physical energy released at the source is a pervasive need of monitoring operations. This goes to the heart of defining the explosion yield levels associated with diverse monitoring thresholds, as well as playing a key role in event identification procedures. Various technologies make different contributions to estimating size. It is difficult to employ radionuclide identification to determine the size of a nuclear test because there are many processes that attenuate the size of various radionuclide signals, such as soil and water absorption, atmospheric fallout, rain-out, delayed releases of radioactive noble gases, and different atmospheric absorption and diffusion processes. Modeling of these effects is still warranted, given that explosion size may have a role in attribution and technology assessment, but radionuclide monitoring plays its most critical role as a discriminant that a nuclear explosion has occurred.

With the other monitoring technologies, size estimation is essential for assessing the levels of national monitoring capabilities. If an event is identified confidently as a nuclear explosion, it is then possible to estimate the yield of the explosion from empirical or theoretical knowledge of how explosion sources generate the observed signals. As described in Section 2.1, the propagation effects for hydroacoustic, infrasound, and seismic waves must be understood if observed signal amplitudes are to be related to source energy release, and this must be done without a priori knowledge of the source type. In effect, there is a need to know the variation of signal amplitudes with distance from a source for different frequencies and wavetypes. This allows the effective source strength to be estimated as a function of frequency and signal type, which provides the basis for many event identification procedures.

Just as for wave travel times, characterizing propagation effects on amplitudes in different media can involve either empirical or model-based strategies. Amplitude-distance curves, analogous to travel time-distance curves, can be used to estimate amplitudes at the source from amplitudes observed at the recording stations. This is a standard practice in seismic magnitude determination. Model-based strategies involve systematic studies of a medium to determine the nature of wave propagation in the medium as represented by one-, two-, or three-dimensional models. This requires techniques that can accurately solve the acoustic or elastic wave propagation equations in the model for the phases of interest. Such approaches estimate the source strength using the entire recorded signal.

Seismic event size or magnitude is based on measuring the amplitude of particular phases on seismograms (further details of magnitude estimations are given in Appendix D). The four most common measures of magnitude are: m_b (measured for the onset of short-period P waves), M_S (for long-period surface waves), M_L (peak short-period motion of S waves at local distances), and $m_b(Lg)$ (short-period surface waves at regional distances). All of these phases are corrected for distance using empirical amplitude-distance curves as discussed in Appendix D. The range of magnitude-yield variations with source emplacement

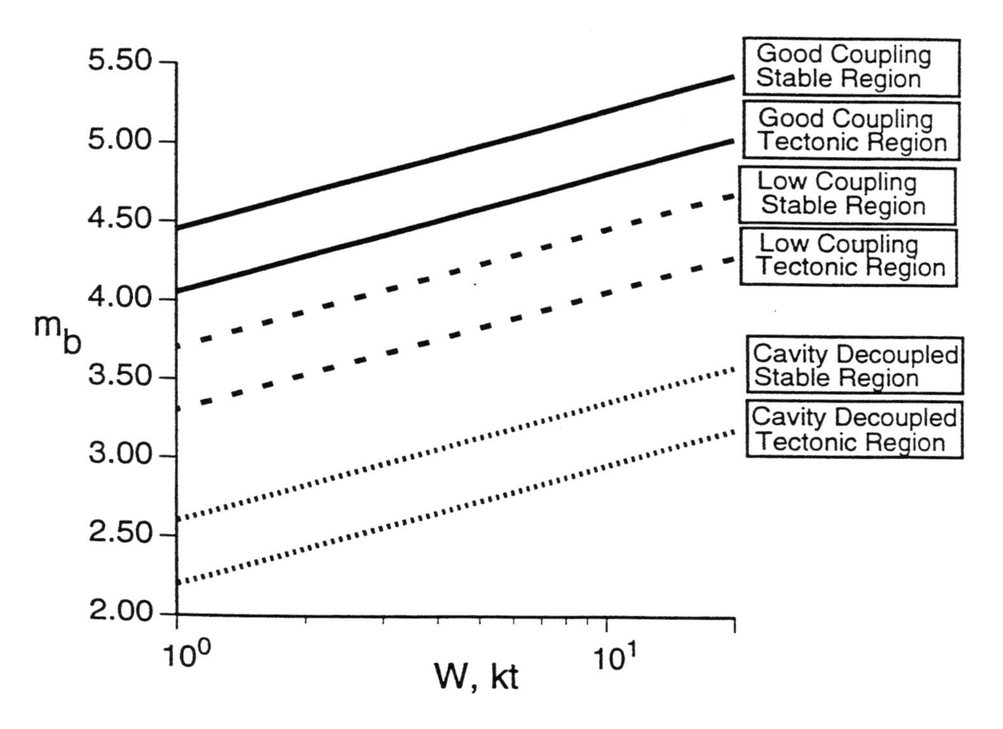

FIGURE 2.2 Comparison of m_b-yield (W) relations for underground nuclear explosions, illustrating the effects of test site tectonic environment and cavity decoupling. SOURCE: Murphy, 1996.

media is illustrated in Figure 2.2. Analogous relationships exist for hydroacoustic and infrasonic measurements as functions of explosion size in the oceans and atmosphere. In the hydroacoustic case, coupling between elastic waves in the solid Earth and acoustic waves in the ocean is sufficiently poorly understood that acoustic amplitudes cannot be used to quantify the size of a source located in the solid Earth.

Effects of Decoupling

One of the most challenging monitoring issues is the reduction of seismic magnitude by decoupling underground explosions in a preexisting cavity. There is a long history of experiment, calculation, and debate about the seismic magnitude effect of decoupling, and this continues to be a critical issue for the CTBT because it forms the basis for one of the more challenging evasion scenarios (e.g., see Sykes, 1996, for a review of this topic). It is widely agreed that full decoupling can achieve a seismic wave amplitude reduction up to about a factor of 70, or 1.8 magnitude units (e.g., Stevens et al., 1991). This means that in a region where a 1.0 kt fully coupled event produces a seismic magnitude of 4.3, one must be able to monitor down to a level of 2.5 with high confidence if decoupling is considered viable and seismological methods are the principal monitoring technology (see Figure 2.2). This capability would provide monitoring down to about 0.01 kt for fully coupled explosions.

If a deep underground nuclear explosion is conducted at the center of a large spherical cavity in hard rock, seismic signals are reduced by a decoupling factor of about 70 if the radius is greater than or equal to about 25 m times the cube root of the yield in kilotons (i.e., $r \geq 25 \times W^{1/3}$).

For a 1 kt explosion (~25 m), this is a size that can be excavated by dissolution mining in salt or by room-pillar mines in hard rock. In fact, much larger solution cavities have been mined in salt, which in principle could decouple explosions with yields as great as several hundred kilotons (the resulting signals would still be large enough to detect and identify), although the largest fully decoupled explosions that have been tested are less than 1 kt. No country is known to have removed the brine from such a dissolution cavity and exploded a device within one, and the distribution of large salt deposits is limited. As noted above, explosion within a fluid-filled cavity can actually enhance the magnitude, so evacuation is essential. Open-air cavities in salt are stable only at depths of about 200 to 1300 m, and the geological distribution of such open-air deposits is also quite limited (Sykes, 1996). Hard rock sites are much more widely distributed; thus, this evasion method must be considered for a number of regions. Explosions can be conducted in cavities produced by prior nuclear explosions as well (all known decoupling tests have been). The prior explosion has to be at least 20 times larger in yield than the fully decoupled event, so the locations of all such cavities relevant to a 1 kt evasion attempt are known (Sykes, 1996). Recent data suggest a smaller decoupling factor of about 10 at high frequencies of 10-20 Hz (Murphy, 1996), which may provide one way to monitor decoupled events with sparse networks. However, these higher frequencies are not likely to be observed at typical regional monitoring distances because of the effects of attenuation.

It is easier to build cavities of the same volume that are elongated rather than spherical, and apparently such aspherical cavities can achieve high decoupling factors, but they also increase the concentration of stress on the cavity and make it much more likely that radionuclides will be released into the atmosphere. An overall evaluation of the cavity decoupling scenario therefore raises several different technical issues:

- Does a country considering an evasive test have access to a suitably remote and controllable region with appropriate geology for cavity construction?
- Can the site be chosen to avoid seismic detection and identification (given that seismic events are reported routinely down to magnitude 2.5 by earthquake monitoring agencies for many areas in industrialized countries)? Can cavities of suitable size, shape, and depth be constructed clandestinely in the chosen region?
- Can nuclear explosions of suitable yield be carried out secretly in sufficient number to support the development of a deployable weapon? Can radionuclides be contained?

Source Identification

The most direct means of nuclear event identification is radionuclide detection. Nuclear power-plant emissions and natural radioactivity (including that from the history of past nuclear testing) constitute the background level against which a nuclear event must be identified. Fortunately, the relative abundance of radioactive isotopes is quite characteristic of the source, and with sufficient signal levels, the type of nuclear device can be determined if measurements are made before critical radionuclides decay. This determination includes the type of fissionable material used (either uranium-235 or plutonium-239) and possibly other nuclear characteristics of the device. Determining the source location for emissions involves complicated atmospheric backtracking and possibly association with detections by other methodologies.

Nuclear event identification can also be based on distinctive features such as the double light flash of atmospheric fireballs if they are observed. For CTBT monitoring purposes, however, the identification process is largely one of determining which events are definitely *not* nuclear tests. For example, accurate determination that the depth of a source in a continental region is greater than 10 km ensures that it is not a human-induced event. A seismic event that is located under the ocean and lacks any hydroacoustic signature of a bubble pulse is deemed to not be a nuclear test. The extent to which the monitoring network accurately can determine critical location parameters, such as depth and offshore location, greatly impacts the number of events that must be identified based on other signal characteristics. There will inevitably be many events that are not identified by straightforward procedures. For these events, more information must be extracted from the signals than is required for any other monitoring procedure, with the result that high-confidence identification thresholds are intrinsically at higher levels than location thresholds. The variability in signals associated with source, path, and receiver effects places identification in a probabilistic context that ultimately involves trade-offs between the confidence levels for not missing events of interest and the estimates of tolerable numbers of mis-identified nonnuclear sources (potentially leading to costly On-Site Inspections or loss of confidence in compliance). As with other monitoring functions, event identification capabilities will vary geographically and with the medium under consideration.

The basis for identifying the type of source for a given wave disturbance (commonly called classification in the hydroacoustic community and discrimination in the seismic and infrasonic communities) involves the distinctive physical processes at the source and the associated excitation of wave energy. For example, the rapid rise times, short source durations, and singular occurrence of underwater explosions and their associated bubble pulses are quite distinctive from submarine volcanic eruptions (which tend to occur in prolonged sequences). The oceanic waveguide formed by the water column and underlying crust causes both explosions and earthquakes to generate long-duration wavetrains. Infrasonic waves excited by an atmospheric or shallow underwater nuclear blast will have frequencies quite distinctive from most longer-duration processes such as volcanic eruptions, severe weather, or meteor impacts. One of the major challenges is thus to characterize not only the wave excitation by explosions but also those of all other sources that can produce signals of size relevant to the monitoring threshold.

For regions in which past nuclear tests were recorded it is possible to estimate the statistical identification capabilities of each monitoring methodology for events in that region. This experience base is actually quite restricted, and identification capabilities established by historical testing in a particular region may not extrapolate to small explosions in the same region or to different regions. Extrapolation to small events is uncertain because the stations available for detection of suitable signals could be different from those used for larger events and distinctive propagation effects can obscure the source identification. Similarly, lateral variations in wave transmission can change the event diagnostic criteria from one region to another by amounts that are difficult to predict based on limited experience with well-calibrated regions. For many regions of the world, CTBT verification will necessarily rely on source identification procedures that are never calibrated directly by recording a nuclear explosion in those particular regions. This places great weight on establishing a fundamental understanding of why identification procedures work when they do, in order to predict their behavior in regions for which only nonnuclear events may be available to characterize the wave behavior and the use of chemical explosions as calibration events. The background levels of nonnuclear activity (natural or human) and the statistical fluctuations in background activity must also be understood to determine the confidence level for the identification of nuclear events and the associated false-alarm potential for nonnuclear events.

The rapid onset time scale and compact source dimensions of underground nuclear explosions have been noted earlier. These directly affect the spectrum of seismic wave energy radiated by the source, with explosions having weak long-period body and surface wave excitation but strong high-frequency radiation. Earthquake faulting typically involves shearing displacements across a fault surface, generating stronger S-wave radiation than P-wave radiation and longer surface waves. Explosion sources are located within the upper kilometer or so of the crust, whereas most earthquakes occur at greater depths.

The depth of the source, the spatial extent of the source, the rate at which source energy is released, and the geometry of the source process all control the relative levels and frequency content of P and S energy released into the surrounding medium (and their partitioning into body and surface waves). Resulting variations in the properties of seismic waves generated by these sources lie at the heart of seismic discrimination methods, as discussed in Appendix D. Comparison of the relative behavior of various measures of source strength tuned to different frequencies or wavetypes (e.g., m_b and M_S), can distinguish the different types of sources in some cases, as illustrated in Figure 2.3. The empirical trends observed in such measures establish a basis for identifying a given source type; these indirect measures must be viewed as statistical indicators.

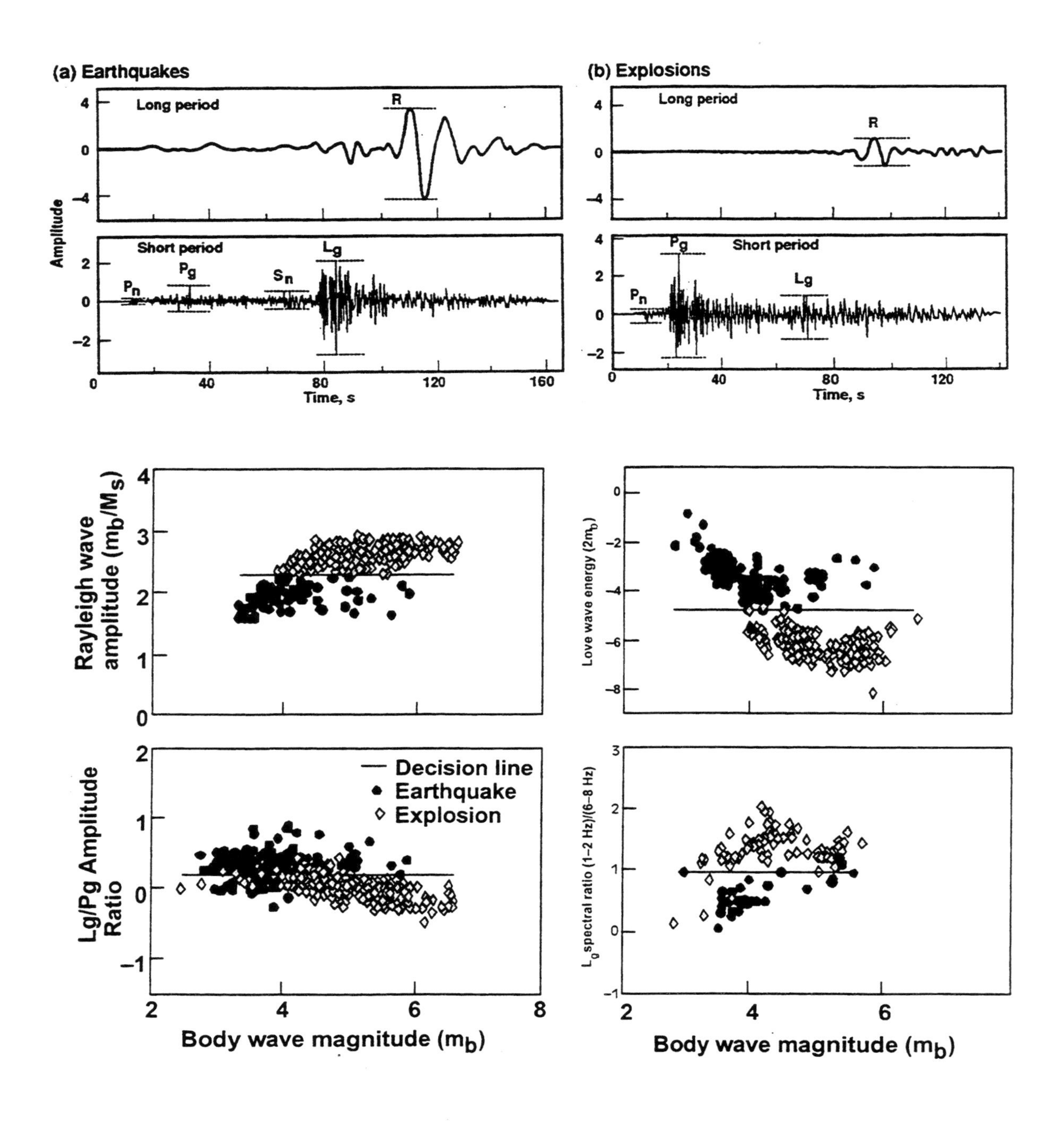

FIGURE 2.3 *Top*: Characteristic seismic signals for explosion and earthquake sources from long- and short-period instruments, illustrating measurements of amplitudes used for magnitude and yield estimates. Pn and Sn are regional P and S waves that travel just under the crust. Pg and Lg are regional P and S waves that propagate within the crust through multiple reflections. *Bottom*: Examples of seismic wave discriminants. SOURCE: DOE, 1993.

For large events, the m_b:M_S comparisons made for teleseismic measurements provide one of the most robust seismic discriminants between explosions and earthquakes. As shown in Figure 2.3, the separation of source types is good for events with body wave magnitudes larger than 4.0. Past research has quantified the empirical success of this discriminant and explained it in terms of the relative excitation efficiency of short-period body wave and 20-second-period surface wave energy by shallow explosions and earthquakes. The confidence gained by achieving a theoretical understanding of this discriminant for the small number of regions with prior nuclear testing allows it to be applied globally for large events. For CTBT monitoring, in which small events are of interest, application of the traditional form of this discriminant is limited by the fact that small events do not excite 20-second-period surface waves, independent of the type of source. Thus, successful teleseismic discriminants developed for large events typically have to be modified for application to regional signals where the propagation effects and the signal frequency content may differ. Appendix D indicates that events with magnitudes down to 3 or lower may be identified based on regional discriminants analogous to the teleseismic discriminants.

In addition to extending successful teleseismic discriminants to the regional environment, the large amplitudes of regional phases can be exploited to develop new seismic discriminants. For example, as shown in Figure 2.2, ratios of regional S-wave-dominated energy to regional P-wave energy (e.g., Lg/Pg), provide some separation of explosion and earthquake observations. However, there is no separation in the magnitude range 3 to 4 in this example. The Lg/Pg ratio has been shown to improve in discriminant performance as the frequency content increases, with frequencies higher than 5 Hz leading to good separation of explosion and earthquake populations in most regions where it has been tested. Improved propagation corrections for lower-frequency signals may enhance the performance of the discriminant as well. Other measures of regional phase energy, emphasizing relative measures of shear and compressional energy or variations with frequency, hold promise for small event identification. Applications to available data sets indicate that regional adjustments in the discriminant baselines and propagation effects have to be determined, effectively involving a calibration effort analogous to that for regional travel time corrections. However, it has also been demonstrated that for reasons that are not well understood at this time, discriminants vary in their performance from region to region, with some doing well in one region of the world but failing to work in others. As a result, there is as yet no single dominant regional seismic discriminant for smaller events that performs as well as m_b:M_S does for larger events. In current practice, then, a suite of regionally calibrated discriminants must serve for small event identification purposes, with sequential application of the discriminant or a statistical combination of the event identification probabilities provided by the various methods being used to define the confidence level for identification for the suite.

The primary obstacles to seismic identification are an inadequate understanding of regional variations in seismic wave propagation, the large variability in signals that constitute the background noise, and the similarity in signals from nuclear and some nonnuclear sources. Although all seismic discriminants are based on empirical measurements that isolate source characteristics, substantial theory has been developed to characterize some successful methods. For monitoring areas with no direct calibration of nuclear explosion signals, one is forced to rely heavily on combined theoretical and empirical validation of the identification methodology, particularly to establish the confidence levels to be assigned to each identification.

For any discriminant measurement there will be scatter in the populations measured for explosions and for other sources. This scatter results from intrinsic measurement error and the variability of the sources (e.g., radiation pattern effects on P and S waves for earthquakes, near-source material properties, heterogeneity along the propagation path, and depth effects on wave excitation). The identification process then involves establishing for a given discriminant a decision line that separates nuclear and nonnuclear events with an acceptable balance between missing a violation and raising a false alarm (Figure 2.4). The overlap of values of the discriminant for the two populations determines the relative probabilities of missed violations and false accusations that can be achieved by the discriminant. Specific values of the two types of error

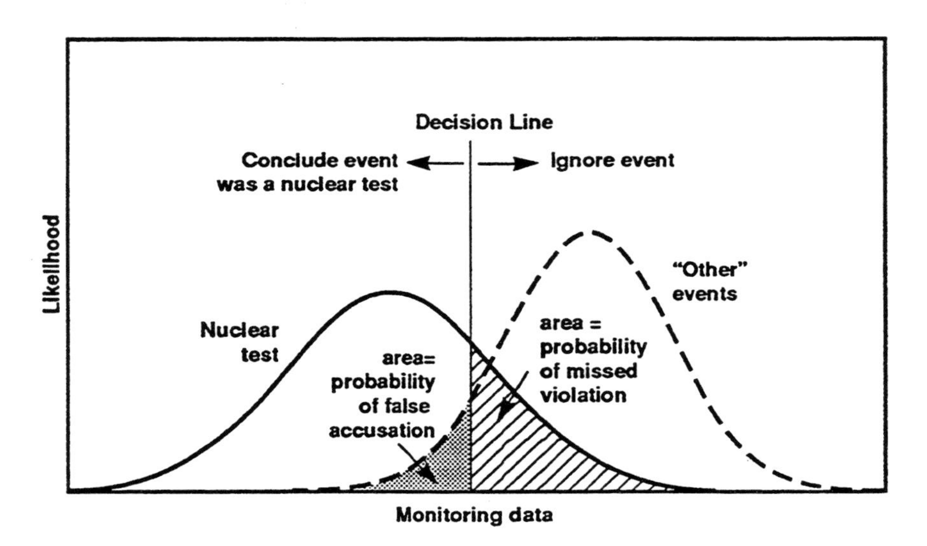

FIGURE 2.4 Schematic illustration of a monitoring decision line. Interpretation of monitoring data is based on values above or below a specified threshold. The decision line determines the probabilities of false accusations and missed violations. SOURCE: DOE, 1993.

area associated with each choice of decision line and the ensemble of values corresponding to various choices are a description of the operating characteristics of the discriminant. Research developments that narrow the overlap of nuclear and nonnuclear distributions are a primary way of improving event identification capabilities. The setting of the decision line is influenced mainly by policy decisions regarding the acceptable tolerance for false alarms versus missed detections. Given the limited experience with applying regional seismic discriminants around the world, event identification for small underground explosions stands out as one of the primary areas for improvement of national monitoring capabilities. Associated research issues are addressed in Chapter 3.

Attribution

If a nuclear test occurs, verification ultimately requires the ability to identify the nation or organization responsible for it. Although various monitoring technologies may indicate high probability that a nuclear device has been detonated at a particular location, on-site, near on-site, or possibly remote-site detection of fission products plays the most important role in identifying the source type as nuclear or nonnuclear. Given that a nuclear explosion has been identified, attribution in the CTBT context can occur in three generic ways either alone or in combination: (1) NTM or intelligence assets identify the nation or organization associated with the explosion; (2) the identified explosion occurs at a location that can be demonstrated to be under the control of a particular nation or organization; and/or (3) analysis of the debris or artifacts from an identified event reveals characteristics that can be associated with a specific nation or organization. The first approach can take place before, during, or after an explosion. It involves sources of information that are outside the scope of this report. The second and third approaches take place after the event has been identified. Figure 2.5 illustrates the generic steps involved in these two approaches.

Attribution on the Basis of Location

In some cases, the site of an identified event may be such that the location alone is sufficient to

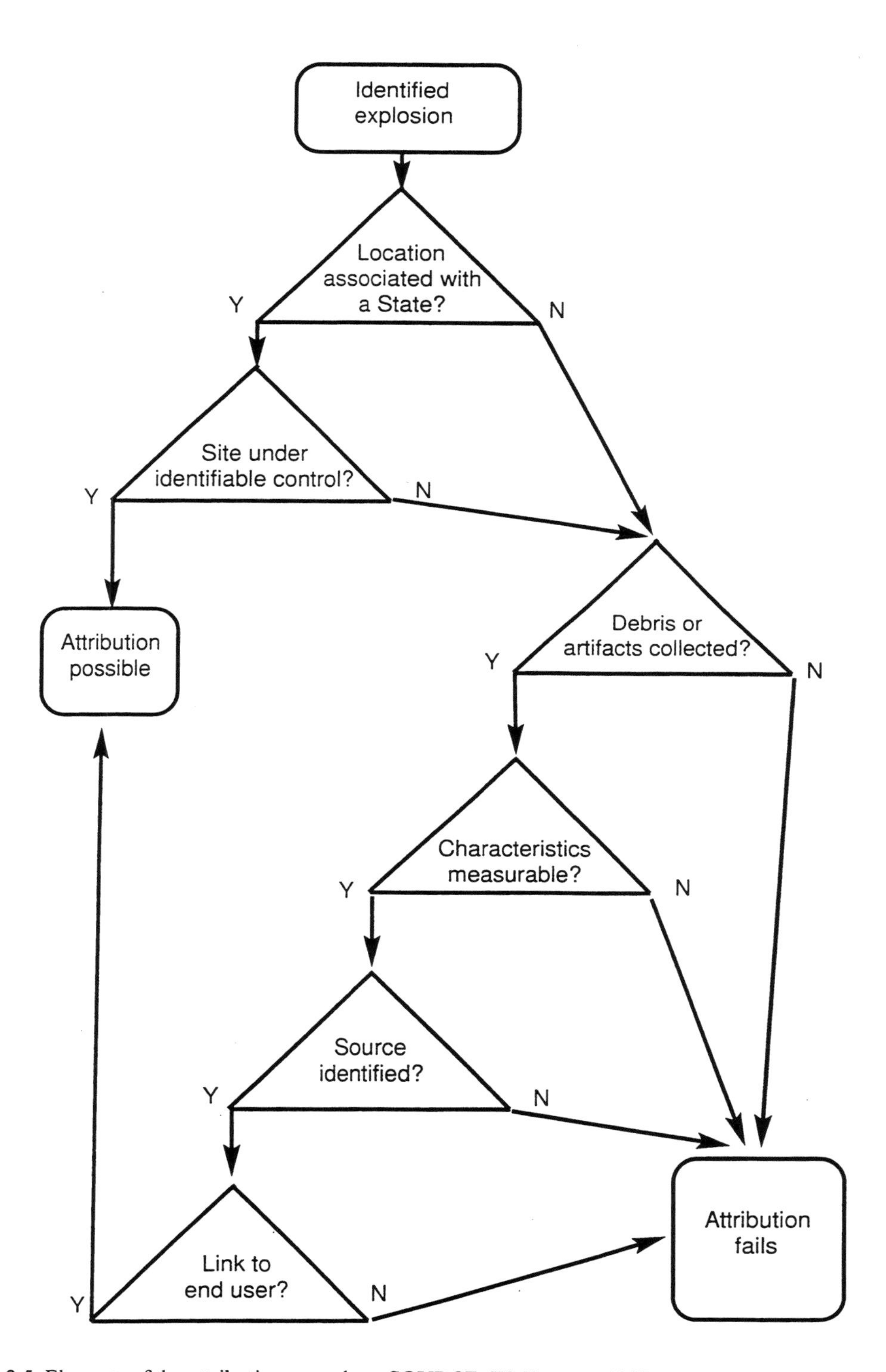

FIGURE 2.5 Elements of the attribution procedure. SOURCE: W. Hannon, 1997.

associate it with a specific nation or organization. In general, such an association would require that the accuracy of the location be sufficient to assign it to a region that was under the unequivocal control of an identifiable state or organization at the time the explosion occurred. Some examples of cases in which these conditions may not be met follow:

- The initial location is near the boundary between two states; the uncertainty in the location is such that it could be on either side of the border; and an On-Site Inspection is unable to locate the site precisely.
- The location is in a region whose actual control is uncertain. Remote areas accessible to multiple parties, areas in dispute, and extended maritime boundaries (e.g., 200-mile limits) are examples of this situation.

Attribution on the Basis of Debris or Artifacts

If attribution on the basis of location is not possible, it may be possible to attribute the event to a specific nation or organization on the basis of debris or artifacts. For this form of attribution to occur,

- debris or artifacts must be recovered,
- appropriate characteristics must be measurable from the debris or artifacts,
- the characteristics either alone or in combination with other information must have distinguishing features that can be correlated with a specific state or organization, (this implies that a reference exists relating specific sources of material or processes to the characteristics), or
- the sources must be tied to an end user.

Debris may be recovered in a variety of ways. In the case of an atmospheric explosion or a massive vent from a shallow underground or underwater explosion, enough material may be present in the atmosphere that remote radionuclide particulate detectors will be able to collect a sample of sufficient size. Airborne collectors, if available, could greatly enhance this process. If such venting does occur, then the near-source fallout will also be detectable by On-Site Inspections or searches at sea, if performed rapidly.

If the event is on land and has not vented, the debris would have to be acquired by drilling. This would require determination of a precise location because samples required for determination of the characteristics can be acquired only from debris located in the cavity beneath the explosion or dispersed through cracks in the underground environment. A well-contained explosion is unlikely to show significant dispersal of material or gases shortly after the explosion. The cavity material would be concentrated in a region with a radius of $10\text{-}30\ m/kt^{1/3}$ (the radius of the cavity -associated with the explosion under various emplacement conditions ranging from fully coupled to fully decoupled). Samples have been recovered under controlled conditions at the U.S. test site when the source's location was known precisely and the drilling was carefully controlled. These conditions are unlikely to exist in the case of an On-Site Inspection and subsurface sample recovery will be correspondingly more difficult.

If the event is located at sea either in the water or in the atmosphere at low altitudes above the surface, it may be possible to obtain samples from the slurry deposited on the surface of the sea from venting or rain-out. Models (T. Harvey, Lawrence Livermore National Laboratory, personal communication, September 1996) indicate that airborne radiation measurements may be able to detect the presence of this slurry as late as two weeks or more after the detonation. These estimates are derived from models incorporating scenario-dependent terms for source and slurry production, long-range dispersion, decay of the radioactive materials, and depletion of the slurry material. Recovery of the material depends on weather conditions and on having an accurate location of the test, both of which have substantial uncertainties. Once the location of the slurry is determined, samples could provide information useful for attribution. For example, if a 1 kt fission explosion occurred on a barge, slurry samples with volumes ranging from a liter to a few hundred liters would be sufficient if they were collected within a week of the detonation. If the sampling time is three weeks later, the required volumes would be larger, but not prohibitively so. The exact volume depends on the nuclide of interest, the particulars of the device (e.g., its yield and special nuclear material), and the evasion scenario.

Analysis of the fission and activation product distribution of recovered particulates may reveal the type of special nuclear material that was used (uranium or plutonium), its preshot isotopic composition, production signatures (e.g., age and trace elements that can be related to the production methods), and/or the materials in structures around the device. If the appropriate data bases are available, these observations might be related to the availability of resources and the capabilities of various nations or organizations. For example, if a radioactive signature revealed that plutonium was the fissile material used in a test, this would limit the source of the device to nations that had the ability to produce or acquire sufficient stockpiles of plutonium to conduct a test.

Even if the properties of the debris can be related to the capabilities and/or resources of a particular nation or organization, attribution of a test on the basis of debris characteristics requires examination of the entire context. For example, the testing nation may have acquired the material from another source. In this case, the source might be in violation of the treaty by encouraging or participating in the test, but it would not have been the testing party.

In principle, each of these steps in an attribution process is possible. The ability to do so will require the development of analytical processes as well as appropriate data bases.

On-Site Inspection

The CTBT provides that "each State Party has the right to request an On-Site Inspection . . . in the territory or in any other place under the jurisdiction or control of any State Party, or in any area beyond the jurisdiction or control of any State." It further states: "The sole purpose of an On-Site Inspection shall be to clarify whether a nuclear weapon test explosion or any other explosion has been carried out . . . and, to the extent possible, to gather any facts which might assist in identifying any possible violator." The Executive Council of the CTBT organization must approve an OSI request by at least 30 affirmative votes.[18]

The protocol to the treaty provides that allowable techniques and equipment may include position finding; visual observation, video and still photography, and multispectral imaging, including infrared measurements; measurements of radioactivity using gamma-radiation monitoring and energy resolution analysis; environmental sampling and analysis of samples; passive seismological monitoring for aftershocks; resonance seismometry and active seismic surveys, magnetic and gravitational field mapping, and ground penetrating radar and electrical conductivity measurements; and drilling to obtain radioactive samples. Overflights are permitted "for the purposes of providing the inspection team with a general orientation of the inspection area, narrowing down and optimizing the locations for ground-based inspection and facilitating the collection of factual evidence."

The approval process and allowance for transportation and site access are such that inspection operations could begin as late as 13 days after the Executive Council receives a request from a State Party. The time elapsed from the occurrence of the event of concern to the start of the inspection could be considerably longer, depending on the internal decision-making processes of the requesting State Party. The inspection team can remain on-site through the twenty-fifth day following the Executive Council's approval of the On-Site Inspection request without further action by the council. The Executive Council may then decide if the inspection is to continue, based on the contents of a report submitted by the inspection team. If the Executive Council does not vote to terminate the inspection at that time, the inspection may continue for up to 35 more days. The inspection may be extended beyond this time by a maximum of 70 days if the team requests an extension and the Executive Council approves it.

The technologies and procedures provided for On-Site Inspections are intended to detect artifacts associated with the conduct of and diagnostics for the explosion, the radioactive evidence left from the explosion, and/or the effects of the explosion on the surrounding media. The artifacts could include any piece of equipment or other physical evidence consistent with nuclear explosion operations. In addition, evidence of recent activity in the area (e.g., roads, drill pads, drill holes, tunnels), although not conclusive evidence of a violation, can serve to focus inspection assets on specific sites. Radioactive evidence includes particulates deposited on the surface by venting, gases that

[18] The Executive Council will consist of 51 members.

escape to the atmosphere by venting or seepage, and residual radioactive material that is trapped in the emplacement medium after the occurrence of an underground or underwater nuclear explosion.

The effects of the explosion on the surrounding material (Figure 2.6) vary depending on material properties, emplacement conditions, and size of the explosion. They range from craters and rubble in chimneys above the point of explosion, to radioactive gases and aftershocks, to effects on the gravity and magnetic fields, the water table, plant life, surface features, and the velocities and attenuation of the seismic waves propagating in the material. For a 1 kt explosion detonated underground, most of these phenomena are concentrated within about one hundred meters of the emplacement point and most are concentrated within the radius of the cavity (roughly 25 m) that would be produced by a well-coupled explosion. (Radioactive materials and gases that reach the surface may be dispersed over a larger region by local winds.)

These effects have different temporal distributions. Collapse craters, if they appear at all, usually do so within an hour but may not appear until long after the occurrence of the explosion. They and any (underground) rubble persist for hundreds or thousands of years. Radioactive materials appear in various time frames. If an underground explosion vents dynamically or the explosion is conducted in the air, gases and particulates may appear on the surface within minutes. If gases reach the surface by traveling along fractures in the rocks, they may take minutes to months to reach levels that are detectable at the surface. They may come to the surface several hundred meters from the explosion by propagating along faults. The significant noble gases have short half-lives (e.g., xenon-131m [12 days], xenon-133m [2 days], xenon-133 [5 days] and xenon-135 [9 hours]) and dissipate in the atmosphere, whereas long-lived radioactive particulates may persist in the vicinity of the explosion for years. Wind, of course, could cause widespread distribution and dilution.

Deformation of the material surrounding the explosion or alteration of its properties in the course of the emplacement might create measurable anomalies. Voids created by the explosion or by the removal of material in the creation of a cavity distort the local gravitational and magnetic fields. Displacement of material and cracking of the rock can affect the water table in the vicinity of the explosion. The displacement of the water table itself may be detectable as may the electric currents generated as the water table returns to an equilibrium state. Distortions of the gravitational and magnetic fields are long-term effects, and displacement of the water table and the return flow can persist for years. In addition, the cavity and any deformation of the material may be detectable by active seismic or electromagnetic probing. It may also be possible to excite a resonant response in the cavity by generating strong seismic signals from surface sources.

If the explosion generates an appreciable shock wave, that wave could interact with the free surface and cause a variety of mechanical (e.g., overturned rocks, cracks, landslides) and biological (e.g., disturbance of the root structure of plants and thus their ability to take up water) effects. These biological effects can persist as long as several months. Their detectability depends in part on weather conditions after the explosion.

The aftershocks that may occur following an explosion depend on many factors, including material properties in the immediate vicinity of the explosion, state of local tectonic stress, and conditions under which the explosion was fired. At one extreme, a fully decoupled explosion may not create any aftershocks or any permanent deformation around the cavity. At the other extreme, a tamped nuclear explosion in a prestressed region will damage the surrounding material and could trigger earthquakes, that relieve local tectonic stress. When aftershocks do occur, the largest ones tend to be 1 to 3 seismic magnitude units smaller than the original explosion and concentrated near the explosion's source region. Their rate of occurrence diminishes with time, and they may eventually become so infrequent as to lose their diagnostic value. For a 1 kt explosion, the estimated time to loss of utility may be as short as 14-21 days.

2.3 MONITORING INFRASTRUCTURE

To perform all of the necessary functions of Section 2.2, the signals collected from IMS sensors must be consolidated in a single data analysis system with little time delay. The concept underlying the IDC is one of continuous transmission

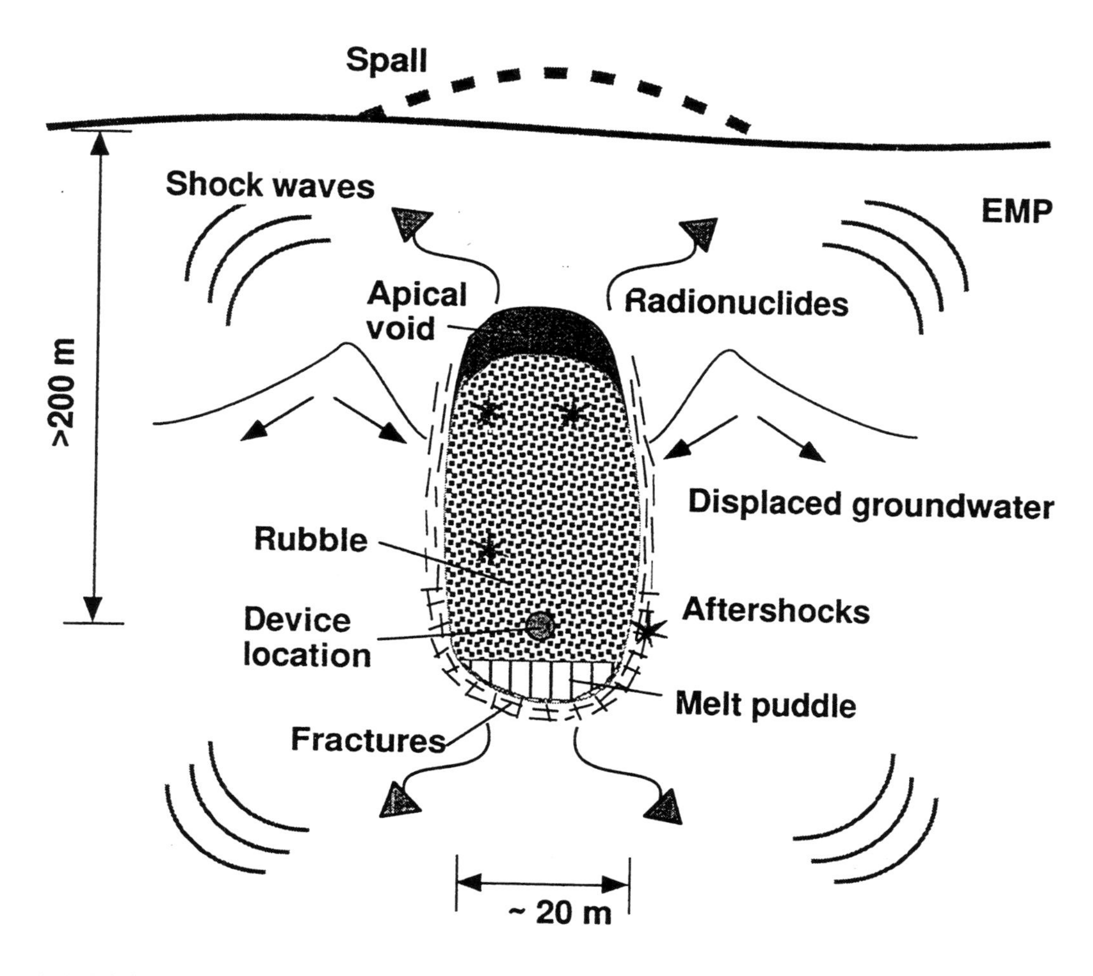

FIGURE 2.6 Near-source environmental effects of an underground explosion. SOURCE: DOE, 1993.

of the time series signals recorded by seismic, infrasonic, and hydroacoustic stations to a single analysis center, which continuously processes the gigabytes of data collected on a daily basis to produce a list of events and event attributes that will serve international treaty monitoring activities. This involves a huge technological challenge for continuous operation of communications and automated signal processing systems, with complex fusion of results from different methodologies. Development of the prototype IDC and the associated trial run of this system during the (GSETT-3) played a key role in establishing the viability of such a global monitoring operation. This capability, largely built on the historical practices of prior seismic monitoring operations by ARPA and AFTAC, as well as operations of seismic arrays around the world, has involved extensive algorithm and computer systems development and implementation. The push for increasing automation of the data analysis has been driven both by the cost consideration of minimizing manpower and by the improved technical capabilities of each generation of computers and software. An attendant motivation is the need for timely event identification in order to initiate On-Site Inspection activities. The longer the delay, the greater are the challenges of detecting ephemeral near-source evidence of a nuclear explosion (e.g., seismic signals from aftershocks).

The U.S. CTBT monitoring effort at AFTAC has incorporated all of the capabilities of the prototype IDC, along with additional procedures for automated event identification (which is not to be performed by the IDC). As the IMS and NTM data sources expand over the next few years, there will be continued need for augmentation of this operational system, and funding to support that activity is essential. In this report, the panel makes a distinction between this type of advanced computer or communication system engineering development and more fundamental research. While this report emphasizes the latter, there is no question that sustained development of both the IDC and the U.S. NDC systems is required, as is allocation of resources for deployment of the actual field data collection systems. The extent to which funding is provided for these activities will strongly influence the time required for CTBT monitoring capabilities to reach the levels described in U.S. goals.

3

Monitoring Technologies: Research Priorities

INTRODUCTION

This chapter summarizes specific research issues confronting the major CTBT monitoring technologies and discusses strategies to enhance national monitoring capabilities through further basic and applied research. The technical challenges and operational issues differ among the technologies, but all of them share the basic functional requirements discussed in Section 2.2. Most of the research needs are discussed with respect to the monitoring challenges of detection, association, location, size estimation, and identification. The panel anticipates that new research challenges and issues will emerge as the IMS is deployed and there is experience with the analysis and use of its data. Presumably these needs will be motivated by "problem" events that defy present identification capabilities. Thus, although this report seeks to identify current areas of research priority, the panel emphasizes that a successful CTBT research program should maintain flexibility to shift emphasis and should nurture basic understanding in related areas that may provide unexpected solutions to future monitoring challenges.

3.1 SEISMOLOGY

Major Technical Issues

The CTBT seismic monitoring system will be challenged by the large number of small events that must be processed on a global basis to provide low-yield threshold monitoring of the underground and underwater environments. Although U.S. monitoring efforts will focus on certain areas of the world, those areas are still extensive. In many cases, the locations lack prior nuclear testing for direct calibration of identification; they require calibration efforts to improve location capability; and, for the most part, they have not been well instrumented seismologically. While historical arrival time observations are available from stations in many regions of the world, readily available waveform signals for determining local structures are less common. Implementation of well-established procedures for calibrating primary functions of the IMS seismic network (in conjunction with additional seismological NTM), such as event detection, association, and location, will provide a predictable level of monitoring capability.

In principle, the stated U.S. goal of high-confidence detection and identification of evasively conducted nuclear explosions of a few kilotons is achievable in limited areas of interest. In practice, doing so will require adequate numbers of appropriately located sensors, sufficient calibration of regional structures, and the development and validation of location and identification algorithms that use regional seismic waves. With the advent of the IMS and planned improvements in U.S. capabilities, many of the current data collection requirements for achieving the current national monitoring objectives will be met. However, additional research is certainly required to use the new data to meet these objectives. Given the current state of knowledge, a number of seismic events within the magnitude range of U.S. monitoring goals would not be distinguishable from nuclear explosions, even if the full IMS-NTM seismic network were in operation. Routine calibration methods will somewhat reduce the upper bound on this population of problem events in certain areas, but even then, research will be essential for significantly improving the overall capabilities of the system. The purpose of the research programs reviewed in this report is to improve monitoring capabilities to the level defined by U.S. monitoring goals.

There are several philosophies in the seismological community about how best to advance the capabilities of seismic monitoring systems, and there is extensive experience with global and regional monitoring of earthquakes and global monitoring of large nuclear explosions. Earthquake monitoring has emphasized collecting data from large numbers of stations, usually in the form of parametric data such as arrival times and amplitudes of seismic phases provided by station operators to a central processing facility. Several thousand global stations contribute data of this type to the production of bulletins and catalogs of the USGS/NEIS and the ISC (see NRC, 1995). Earthquake studies have prompted the development of many global and regional seismic velocity models for use in event location procedures. Many regional seismographic networks process short-period digital seismic waveforms for local earthquake bulletin preparation, and there has been some progress in use of near-real-time digital seismic data for production of the USGS/NEIS bulletins. For these bulletins, the need for prompt publication is usually less that that associated with nuclear test monitoring. When there is a need for a rapid result, as in documenting the location of an earthquake disaster to assist in emergency planning, the USGS/NEIS can and does provide a preliminary location within a few minutes of the arrival of the seismic waves to distant stations. These earthquake-related activities will continue in parallel with CTBT monitoring.

In recent years, global and regional broadband networks deployed by universities and the USGS for studying and monitoring earthquakes have developed entirely new analytical approaches, including systematic quantification of earthquake fault geometry and energy release based on analysis of waveforms. Of greatest relevance to CTBT monitoring are the quantitative approaches for event location and characterization being developed for analysis of seismic signals from small nearby earthquakes. When adequate crustal structures and seismic wave synthesis methods are available, it is possible to model complete broadband ground motions for regional events, enabling accurate source depth determination, event location and characterization, and development of waveform catalogs for efficient processing of future events (see Appendix D). The modeling may include inversion for the source moment tensor[19] Efforts of this type require complete understanding of the nature of all ground motions recorded by the monitoring network.

One of the core philosophical issues for seismic monitoring operations is whether it is better to use global and/or regional travel time curves, possibly with station or source region corrections, or to explicitly use models of the Earth's velocity structure and calculate the travel times and amplitudes for each source-station pair. The velocity models, which can include variable crustal and lithospheric structure, can be derived from the same data used in defining local travel time curves, but once they are determined they could also be used to model additional seismic signals that are not employed in standard event processing, such as free oscillations, surface waves, and multiple-body

[19] The moment tensor is a representation of the set of equivalent forces at the source that would produce the observed ground motion. The sizes and orientation of the equivalent forces are distinctive for earthquakes and explosions and thus form a potential discriminant.

wave reflections. Velocity models are constantly improving on both global and regional scales and provide better approximations to the Earth with each model generation, with corresponding improvements in event location. Velocity models also have a key advantage completely lacking in travel time curves: they provide the basis for synthesizing the seismic motions expected for a specific path, as involved in the regional wave modeling mentioned above. The synthetic ground motions are useful for estimates of improving the source depth, identifying blockage of certain phase types, and enhancing the identification of the source type. The use of travel time curves has been adequate for teleseismic monitoring of large events but may be too limited for dealing with the regional monitoring required for small events.

The nature of the Earth's velocity structure is such that heterogeneity exists on all scales (Figure 3.1), and at some level, interpolation of empirical travel time corrections as well as the intrinsic interpolation involved in aspherical model construction will fail to account for actual effects of Earth structure. For the teleseismic monitoring approach, in which the major phases of interest travel down into the mantle (or are relatively long-period surface waves that average over shallow structure), there is no obvious advantage of using travel time corrections versus three-dimensional models in regions with many sources, other than perhaps the operational simplicity of the former. However, for regions with sparse source or station distributions, the velocity model can incorporate information from many independent wavetypes and paths and predict structural effects for other paths and wavetypes that cannot otherwise be calibrated directly. The confidence gained from correctly predicting the energy partitioning in the seismic signal on a given path by waveform modeling directly enhances the source identification. The value of this approach is not a controversial notion, but it is in tension with the magnitude of resources that must be invested to adequately determine the structure in extensive areas of the world.

When the entire field of seismology is considered, it is clear that the science is moving toward a three-dimensional parameterized model of Earth's material properties that will provide quite accurate predictions of seismic wave travel times for many applications (including earthquake monitoring and basic studies of Earth's composition and dynamics). One component of a long-term CTBT monitoring research program could involve commitment of resources toward the development of an improved global three-dimensional velocity model beginning with regions of interest, perhaps in partnership with the National Science Foundation (NSF; see Appendix C). Previous nuclear test monitoring research programs have supported development of reference Earth models. The operational system could be positioned for systematically updating the reference model used for locations by adopting a current three-dimensional model at this time, possibly including three-dimensional ray-tracing capabilities that will become essential as resolution of the model improves. This approach would provide a framework for including the somewhat more focused efforts of the CTBT research program, such as the pursuit of a detailed model of the crust and lithosphere in Eurasia and the Middle East, for event location and identification. Partnership with other agencies and organizations pursuing related efforts could lead to rapid progress on this goal. Another major coordinated effort could be the development of regional event bulletins complete down to a low-magnitude level, such as 2.5 (achieving a global bulletin at this level would require significant enhancement of global seismic monitoring capabilities). This would require extensive coordination between the earthquake monitoring research and the CTBT monitoring communities, but it is technically viable and would provide a basis for CTBT monitoring with high confidence. Related research issues are summarized below as specific functions of the monitoring system are considered.

Detection

Figure 3.2 shows the projected detection threshold of the 50-station IMS Primary Network when fully deployed (Claassen, 1996). This calculation is based on a criterion of three or more stations detecting P arrivals with a 99 per cent probability. Where available, actual station spectral noise statistics were used, but for stations that do not yet exist the noise levels were assumed to be those of low-noise stations. NTM will enhance the performance of the national monitoring system relative to these calculations in several areas of the

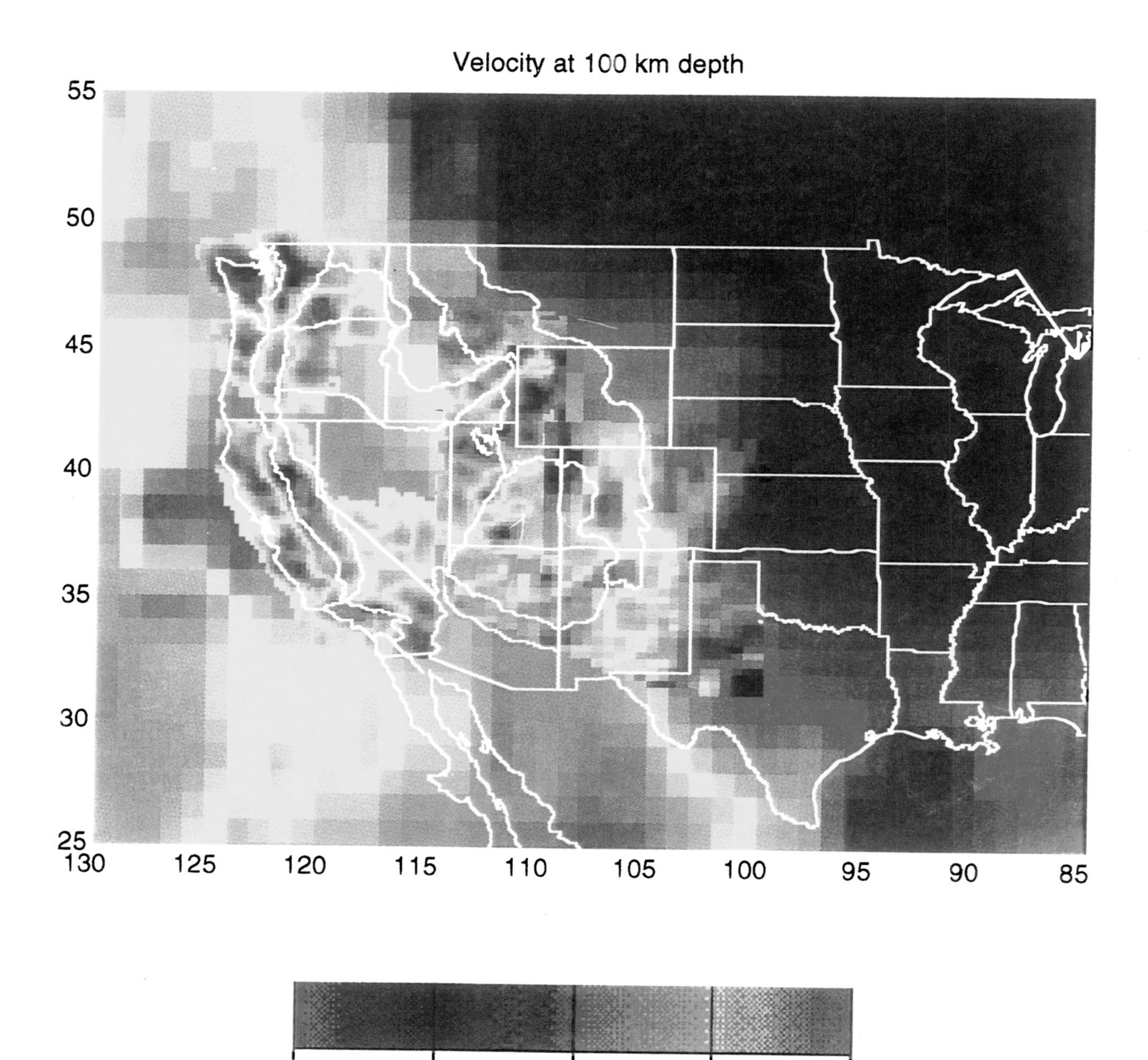

FIGURE 3.1 Heterogeneous S wave velocity variations at a depth of 100 km beneath the western U.S. The scale bar shows the relative velocity variations in per cent. (SOURCE: K. G. Duekers and E. Humphreys, personal communication, 1997).

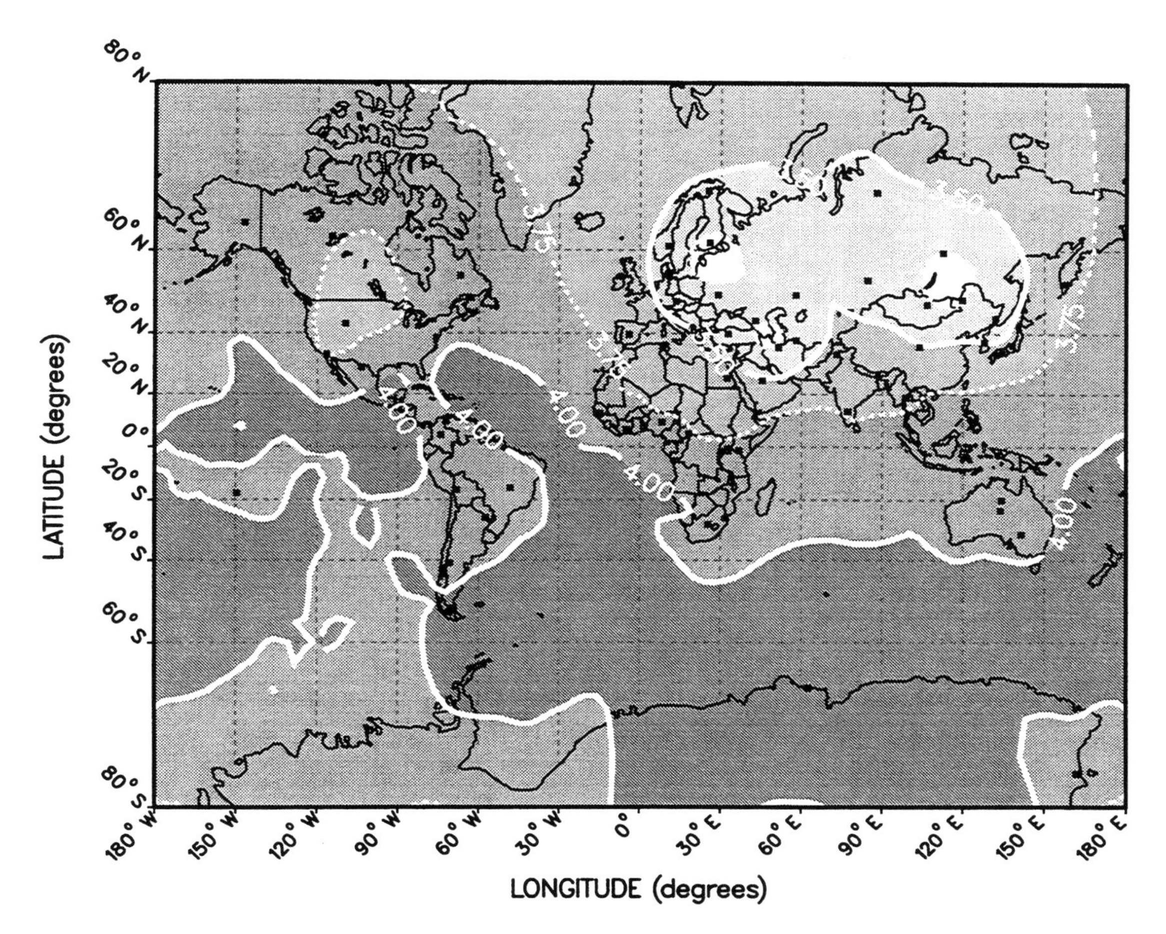

FIGURE 3.2 Detection threshold predicted for the fully deployed IMS Primary Network. The station locations are indicated as black squares on the map. SOURCE: (Claassen, 1996)

world. This simulation, which will have to be validated by actual operations, suggests that IMS detection thresholds across Eurasia will be at or below magnitude 3.75, with some areas being as low as 3.25.

Given the small percentage (5-6 per cent) of detections ultimately associated with events in the first few years of the prototype-IDC REB, and the fact that 30 per cent of the detections in the final event list had to be added by analysts, there are clearly research issues related to improved seismic detection. Original goals for the IDC involved association of 10 and 30 per cent of detected phases from the Primary and Auxiliary stations, respectively. Many unassociated detections are actually signals from tiny local events that occur only at one station. It would not be useful to locate such small events in most regions. It may be possible to screen out such signals using the waveshape information provided by templates of past events recorded by each station. This could reduce false associations and unburden the overall algorithms for association. Three-component stations have a lower proportion of associated detections (4 per cent) than do arrays (6 per cent) in the prototype system, and further research on combining and adjusting automated detection parameters holds promise of improving the performance.

The significant number of detections added or revised by analysts suggests room for improving detection algorithms that run automatically.

Research on enhancing the signal-to-noise ratio and improving the onset time determination has particular value. Exploration of simultaneous use of multiple detectors at a given station may result in approaches to reduce spurious detections and improve onset determinations. Practice at the prototype IDC has found that 76 per cent of the automatic detections have an onset time within 1.0 second of that picked by an analyst, whereas the goal of the IDC is 90 per cent.

A key detection issue is improving how overlapping signals are handled. This includes the problems of both multiple events and multiple arrivals from events. Evasion scenarios that involve masking an explosion in an earthquake or quarry blast require detection of superimposed signals. Time series analysis procedures for separating closely spaced overlapping signals with slightly different frequency content potentially can improve detection in such cases.

Perhaps the greatest room for research progress on detections involves phase identification. Only 7 per cent of teleseismic phases at Primary Network arrays were mis-identified as regional phases by the prototype IDC in December 1995, but 32 per cent of teleseismic phases from three-component stations were mis-identified. The complementary numbers of mis-identified regional phases were 8 per cent for arrays and 48 per cent (P waves) and 27 per cent (S phases) for three-component stations. Improved polarization analysis for three-component stations is needed. There is also a need for improved slowness and azimuth determinations. In the prototype system, azimuth and slowness measurements currently make up 9 per cent and 5 per cent, respectively, of the total defining parameters used in the REB, and the introduction of methods to improve these parameters will enhance both association and location procedures significantly.

There is relatively little experience with detectors for T-phases observed at island seismic stations, and effective algorithms must be developed for these noisy environments. Another area of research is the identification of seismoacoustic detections generated by propagating air waves. Although they are generated by acoustic waves in the atmosphere, they are often called "lonesome" Lg waves because they appear as an Lg phase long after the first arrival. Consequently, they are interpreted incorrectly as a seismic signal to be associated with records from other seismic stations. Analysis of colocated seismic and infrasonic sensors holds promise for solving this problem.

Association

Association is an area in which recent basic research has enhanced CTBT monitoring capabilities, as newly developed Generalized Association (GA) algorithms (see discussion of association in Chapter 2) are being incorporated into routine operations of the U.S. NDC (and the prototype IDC). Active research is under way on incorporating additional information into GA methods, and several approaches have been explored for using relative phase amplitude information and other arrival properties. Use of complete waveforms is also promising, and such methods may find particular value for complex overlapping sequences of aftershocks or quarry blasts. Given the archive of waveform data for older events that will accumulate at the U.S. NDC, innovative use of previous signals as templates and master events should be explored. This is a new arena of monitoring operations, and only limited research has been conducted on such approaches. It is likely that the number of unassociated detections can be reduced by preliminary waveshape screening to recognize local events detected at only one sensor and to remove their signals from further association processing. Incorporation of regional propagation information into the GA, such as blockage patterns for a given candidate source region, using a computer knowledge base should enhance association methods, but this strategy requires further development.

Location

Seismic event location is a key area for further research efforts because accurate locations are essential for event identification and On-Site Inspections. Formal procedures exist for assessing the precision in event location (see Appendix C), but these typically do not account for possible systematic error and hence may overestimate event location accuracy. Figure 3.3 presents a

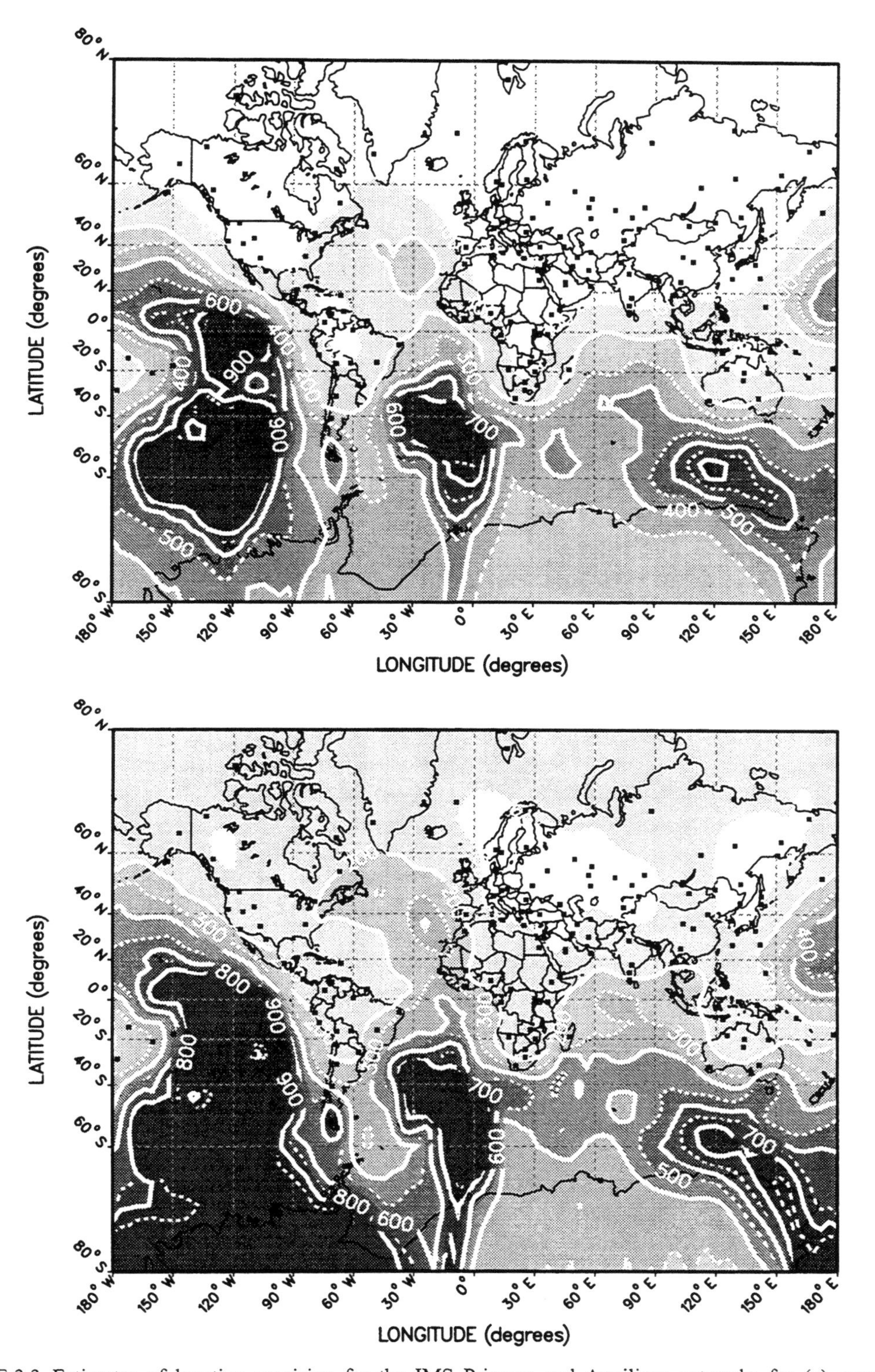

FIGURE 3.3 Estimates of location precision for the IMS Primary and Auxiliary networks for (a) magnitude 4.25 events and (b) events at the detection threshold in Figure 3.2. Neither plot includes an appraisal of uncertainty due to systematic error. The seismic stations are indicated as black squares on the map. SOURCE: (Claassen, 1996)

calculation of the expected distribution of the precision of seismic event locations for the fully deployed Primary and Auxiliary networks of the IMS (Claassen, 1996) for events with magnitudes of 4.25 at the detection threshold calculated for the Primary Network (Figure 3.2). As in the case of detection-level calculations, actual station noise levels were used when available, low noise levels were assumed for future Primary Network stations, and low noise levels raised by 10 dB were used for future Auxiliary Network stations. The precision of the location performances is stated in terms of 90 per cent confidence of being within a given area measured in square kilometers. Only P-wave arrival times and azimuths were used, and no extensive calibration was assumed. (Note that this error estimate does not include the uncertainty due to possible systematic error.) Location precision better than 1000 km^2 can be attained in most regions of the world for events at and above the detection level, but these results can be misleading because it is important to allow for biases due to systematic error, which can be substantial. For example, during assessment of the first 20 months of the REB, comparison by various countries of the locations produced by their own denser national networks with those in the REB showed that the REB 90 per cent confidence ellipses contained the national network location (which is presumably more accurate) less than half of the time. It has been estimated that the location precision for events in the REB can be improved by a factor of 6 after calibration of the network.

Two main research strategies exist for reducing systematic error in event locations: (1) the development of regionalized travel times with reliable path calibrations for events of known location, and (2) the development of improved three-dimensional velocity models that give less biased locations. (Appendix C discusses this issue at length.) A catalog of calibration events can advance the latter strategy as well, given the difficulty of eliminating trade-offs between locations and heterogeneity in developing three-dimensional models. From the CTBT operational perspective it is not clear that even the next generation of three-dimensional models (or regionalized travel time curves) will account for the Earth's heterogeneity sufficiently to eliminate systematic location errors for either teleseismic or regional observations. Thus, even if improved velocity models are used, it is desirable to calibrate station and source corrections to account for unmodeled effects.

Empirical calibration of location capabilities for a seismic network can proceed on various scales, ranging from use of a few key calibration events such as past nuclear explosions with well-known locations, to more ambitious undertakings such as development of large catalogs of events including well located earthquakes. Some efforts along these lines are being pursued in the DOE CTBT research program. The challenge is to obtain sufficient numbers of ground truth events with well-constrained parameters. Appendix C considers a systems approach to this problem, motivated by the need for calibration of extensive regions. Quarry blasts, earthquake ruptures that visibly break the surface, aftershocks with accurate locations determined from local deployments of portable arrays, and events located by dense local seismic networks can be used for this purpose because such event locations are in some cases accurate to within 1 to 2 km. There are large uncertainties in how to interpolate calibration information from discrete source-receiver combinations, and research on the statistical nature of heterogeneity may provide guidance in this process. Such uncertainty also serves as motivation for further development of velocity models because the many data sources that can be incorporated into such models can resolve heterogeneities that are not calibrated directly by ground truth events.

A major basic and applied research effort of great importance for event location capabilities involves developing regionalized travel time curves and velocity models, particularly for the crustal phases that will be detected for small events. These regionalized models must be merged with global structures to permit simultaneous locations with regional and teleseismic signals. Determining regionalized travel time curves usually involves a combination of empirical measurements and modeling efforts, with the latter being important to ensure appropriate identification of regional phases (variations in crustal thickness and source radiation can lead to confusion about which phase is being measured). Historical data bases may be valuable for determining regional travel time curves in areas where new stations are being deployed. A demonstrated research tool that assists in this effort involves systematic modeling of regional broadband seismograms for earthquakes and quarry blasts. Confidence in the adequacy of a given regional velocity structure or the associated empirical travel time curve is greatly enhanced if computer simulations demonstrate that the structural model actually accounts for the timing and relative amplitude of phases in the seismogram. By systematic modeling of broadband waveforms from larger earthquakes (for which the source can be determined by analysis of signals from multiple stations), regional Earth structures with good predictive capabilities can be determined. It is then possible either to use the velocity model to predict the times of regional phases or to use travel time curves for which the model has validated the identification of phases. This is of particular value in regions where the crustal heterogeneity (e.g., near continental margins and in mountain belts) causes the energy partitioning to change among phases; for example, some crustal paths do not allow the Lg phase to propagate, and anomalously large Sn phases may be observed instead.

Waveform modeling approaches can play a major role in determining the local velocity structures required to interpret regional phases for both event location and event identification, so there is value in further development of seismological modeling techniques that can compute synthetic ground motions for complex models. Nuclear test monitoring research programs have long supported basic development of seismic modeling capabilities because they underlie the quantification of most seismic monitoring methods. Present challenges include modeling of regional distance (to 1000 km) high-frequency crustal phases (up to 10 Hz and higher) for paths in two- and three-dimensional models of the crust. Such modeling must correctly include surface and internal boundaries that are rough, as well as both large- and small-scale volumetric heterogeneities. Current capabilities are limited, and in only a handful of situations have regional waveform complexities been quantified adequately by synthetics. New modeling approaches and faster computer technologies will be required to achieve the level of seismogram quantification for shorter-period regional seismic waves that now exists for global observations of longer-period seismic waves. In parallel with the development of new modeling capabilities is a need for improved strategies for determining characteristics of the crust based on sparse regional observations, so that realistic velocity models can be developed rather than ad hoc structures.

Other promising areas for research include methods of using complete waveform information to locate events and improved use of long-period energy. Correlations of waveforms can provide accurate relative locations for similar sources such as mining explosions. The basic idea is that rather than relying on only the relative arrival times of direct P, one can use relative arrival times of all phases in the seismogram to constrain the relative location. The potential for regional event location based on such approaches is not yet fully established, but preliminary work with waveforms from mining areas is promising. There is also potential for improving event locations using surface waves because there have been significant advances in the global maps of phase velocity heterogeneity affecting these phases. Because surface waves with periods of 5-20 seconds are valuable for estimation of source size and event identification, signal processing procedures that accurately time these arrivals (notably using phase-match filters that enhance signal-to-noise by correcting for the systematic dispersive nature of surface waves) have to be developed and will provide information useful for locating events. For applications to small events, existing phase velocity models must be improved and extended to shorter periods.

A particularly important aspect of event location requiring further research is determination of depth for small, regionally recorded events. This parameter is of great value for identifying the source but is one of the most challenging problems for regional event monitoring. Based on experience with earthquake monitoring in densely instrumented regions of the world, accurate depth determination is not typically achievable using direct body wave arrival times alone, and more complete waveform information must be used. The complex set of reverberations that exist at regional distances makes identification of discrete seismic "depth phases" difficult, but complete waveform modeling, as well as cepstral methods (involving analysis of the spectrum of the signal) applied to entire sets of body wave arrivals, hold potential for identifying such phases. Research advances in this area potentially can eliminate many events from further analysis in the CTBT monitoring system.

Event location techniques that exploit synergy between various monitoring technologies are described in Section 3.6.

Size Estimation

Every event located by the IMS and NTM will have some recorded signals that can be used to estimate the strength of the source for various frequencies and wavetypes. Because the actual seismograms are retrieved from the field, it is possible to measure a variety of waveform characteristics to characterize the source strength. The standard seismic magnitude scales (see Appendix D) for short-period P waves (m_b) and 20-second-period surface waves (M_S) are the principal teleseismic size estimators used for event identification and yield estimation for larger events. In December 1995, 88 per cent of the events in the REB were assigned body wave magnitudes, but only 10 per cent have surface wave magnitudes because of low surface wave amplitudes for small events and the sparseness of the network. Operational experience at the prototype IDC is establishing systematic station corrections (average deviations from well-determined mean event magnitudes) that can be applied to reduce biases in the measurements, particularly for small events with few recordings. About 34 per cent of the REB events for December 1995 were assigned a "local magnitude," which was scaled relative to m_b.

In addition to station corrections, which account for systematic station-dependent biases, it is important to determine regionalized amplitude-distance curves analogous to regional travel time curves. These curves are used in the magnitude formulations and have great variations at regional distances (see Appendix D). Although range- and azimuth-dependent station corrections can absorb regional patterns, it is valuable to have a suitable regional structure for interpolation of general trends. Research is needed to establish the level of regionalization required and the nature of the regional amplitude-distance curves. As new IMS and NTM stations are deployed, each region must be calibrated and an understanding attained of the nature of the seismic wave propagation in that region (including effects such as blockage, which may prevent measurement of some phases).

The availability of complete waveform information for each event offers potential source strength estimation that exploits more of the signal than conventional seismic magnitudes. For example, complete waveform modeling can determine accurate seismic moments (measures of the overall fault energy release) that may be superior to seismic magnitudes because they explicitly account for fault geometry. Routine inversion of complete ground motion recordings for seismic moment and fault geometry is now conducted for all events with magnitudes larger than about 5.0 around the world by the academic community that is studying earthquake processes. Similar capabilities have been demonstrated for events with magnitudes as small as 3.5 in well-instrumented seismogenic areas (see Appendix D). These approaches are closely linked to source identification because they explicitly incorporate and solve for generalized representations of forces exerted at the source. The extent to which a sparse network such as the IMS can exploit such waveform modeling approaches is not fully established. In part, it depends on the extent to which adequate regional velocity models are determined and on their waveform prediction capabilities. Further research can establish the operational role of complete waveform analysis for source strength estimation.

Operationally, it is usually convenient to use parametric measurements such as magnitudes. Several promising waveform measurements for regional phases can be treated parametrically. These include waveform energy measurements for short- and long-period passbands; coda magnitudes based on frequency-dependent variations of reverberations following principal seismic arrivals; and signal power measurements of Sn, Lg, and other reverberative phases at regional distances. Extension of surface wave measurements to the short-period signals (5-15 seconds) that dominate regional recordings of small events is also necessary. Research on the utility, stability, and regional variability of these source strength estimators should continue. This includes necessary efforts to characterize the effects of source depth, distance, attenuation, heterogeneous crustal structure, and recording site for each approach. This effort is warranted given the limitations of conventional

m_b and M_S measurements for small events recorded by sparse networks.

Identification

Several major research areas related to seismic source identification have been considered in the preceding discussion of event location and strength estimation. Accurate event location is essential for identification, including the reliable separation of onshore and offshore events and the determination of source depth. When location alone is insufficient to identify the source, secondary waveform attributes must be relied on. For events larger than about magnitude 4.5, well-tested methods based on teleseismic data provide reliable discrimination. For small events, experience has shown that a number of ad hoc methods, often different in different regions, can be used to distinguish explosions (e.g., mine blasting) from earthquakes. However, a primary area for both basic and applied research is to systematize such experience for identifying small events and turn it to solving the problems of CTBT monitoring. Some key areas include the following:

- Extension of m_b:M_S-type discriminants to regional scales for small events: this involves development of regional surrogates for both P-wave and surface-wave magnitudes that retain the frequency-source depth-source mechanism sensitivity of teleseismic discriminants. Improved methods of measuring the surface wave source strength for regional signals are necessary, as mentioned above. Quantification of the regional measures by waveform modeling and source theory is needed to provide a solid physical understanding of such empirical discriminants.
- Regional S/P type measurements (e.g., Lg/Pn, Lg/Pg, Sn/Pg) have been shown to discriminate source types well at frequencies higher than 3-5 Hz. Research is needed to establish the regional variability of such measurements and to reduce the scatter in earthquake populations. Improved path corrections, beyond standard distance curves, that account for regional crustal variability should be developed further because they appear to reduce scatter for frequencies lower than 3 Hz.
- Quarry blasts and mining events (explosions, roof collapses, rockbursts) can pose major identification challenges, and research is needed to establish the variability of these sources and the performance of proposed discriminants in a variety of areas. Ripple-fired explosions (with a series of spatially distributed and temporally lagged charges) can often be discriminated from other explosions and earthquakes by the presence of discrete frequencies associated with shot time separation. The discriminant appears to be broadly applicable, although additional testing in new environments is essential. However, it does not preclude a scenario in which the mining explosion masks a nuclear test (see below).
- Systematic, complete waveform inversion for source type should be explored for regions with well-determined crustal structures, given the constraints of the large spacing between seismic stations in the IMS and NTM. This can contribute directly to source depth determination and source identification. It is likely that such approaches are the key to solving challenges posed by some evasion scenarios involving masking of nuclear test signals by simultaneous quarry blasts, rockbursts, or earthquakes.
- Strategies for calibrating discriminants in various regions must be established, with procedures to correct for regional path effects being expanded. This includes systematic mapping of blockage effects and attenuation structure. It is also desirable to establish populations of quarry blast signals for each region.

Summary of Research Priorities Associated with Seismic Monitoring

In summary, a prioritized list of research topics in seismology that would enhance CTBT monitoring capabilities includes:

1) Improved characterization and modeling of regional seismic wave propagation in diverse regions of the world.
2) Improved capabilities to detect, locate, and identify small events using sparsely distributed seismic arrays.
3) Theoretical and observational investigations of the full range of seismic sources.
4) Development of high-resolution velocity models for regions of monitoring concern.

3.2 HYDROACOUSTICS

Major Technical Issues

The ocean is a remarkably efficient medium for the transmission of sound energy. Its deep sound channel (the SOFAR channel; see Appendix E) allows sound energy to propagate with little attenuation over global distances. Even low-level sounds, on the scale of those produced by several kilograms of TNT, can often be detected at ranges of many thousands of kilometers. Despite these attributes, the hydrophone and T-phase station network proposed for the IMS will be inadequate for accurately locating sources in all the world's oceans without ancillary data from other monitoring technologies. In certain cases, such as low-altitude explosions above the sea surface, the hydroacoustic system by itself might be incapable of providing a detection altogether or could be jammed easily. Thus, it is especially important to be able to fuse hydroacoustic data with other IMS data—seismic and infrasonic.

Figure 3.4 estimates the location accuracy of the IMS hydrophone network together with the ocean-island seismic stations for a 1-kiloton explosion 50 m beneath the sea surface. Model simulations show that for some areas, especially regions of high shipping density, location uncertainties can exceed 1000 km^2 by a substantial margin. This is a result of at least two factors: (1) locally high background noise associated with shipping masks the explosive signal, and (2) at least three independent observations of acoustic arrivals are needed to locate a source by triangulation. Given blockage by islands, seamounts, continents, and other features, this is difficult to achieve worldwide with only six hydroacoustic systems and five T-phase stations. None of the IMS hydroacoustic stations have direction resolving capabilities; if available, these would provide improved location accuracy with detections on fewer sensors. In practice, the hydroacoustic data will be examined for discrimination purposes and combined with seismic data for location purposes.

If the detonation is above the sea surface, signal levels at those stations that are not blocked by bathymetry will be substantially lower, and location uncertainties will increase. For example, calculations show that an atmospheric explosion of 1 kt at a height 1 km above the surface will couple energy into the SOFAR channel roughly equivalent to the detonation of a 10-50 kg explosion at channel depth, which is five orders of magnitude less than 1 kt. Sound pressure level scaling (which is proportional to pressure squared) for TNT charges based on empirical data is roughly proportional to ($W^{2/3}$), where W is the weight of the charge. Thus, each factor of 10 decrease in charge weight reduces the effective sound pressure level by about 6.7 dB (in practice, the factor is 7-7.5 dB), and five orders of magnitude decreases the effective source level by about 35 dB.

A reduction in signal-to-noise ratio of this magnitude implies that at certain receivers, signal levels will fall below the background noise and will not be detected at all. In addition, the reduction in signal-to-noise ratio will also reduce travel time accuracies by a factor of about 50 because the variance in measuring travel times is inversely proportional to the square root of the signal-to-noise ratio.

Even if the reduced signal from a low-level atmospheric explosion is above the background noise at a receiver, its level is comparable to many natural (e.g., volcanoes, earthquakes, microseisms generated by waves) and man-made (e.g., seismic profiling) sounds. The signals from such an explosion are, therefore, susceptible to natural noise masking and fairly simple jamming techniques.

The hydroacoustic system alone may also be incapable of detecting subsurface explosions in certain regions where bathymetric blocking exists and where the coupling of energy into the SOFAR channel is poor. In other areas, detection may be possible but localization could be problematic at best. An example might be an explosion in the Antarctic Ocean where the axis of the SOFAR channel at the surface causes all sound energy to

FIGURE 3.4 (Facing Page) (a, top) Estimated location precision for the IMS hydroacoustic network (open circles) combined with the ocean-island seismic stations (triangles) for a 1 kt explosion 50 m below the ocean surface. Station locations are also shown in Figure 1.1a. The scale shows the uncertainty in square kilometers. The linear features of low resolution are associated with regions of high shipping density. (b, bottom) Estimated location precision for the southern oceans. The image on the left is for the combined hydroacoustic and ocean-island seismic network. The image on the right shows the performance for only the hydroacoustic system. SOURCE: D.B. Harris, Lawrence Livermore National Laboratory, personal communication, 1997.

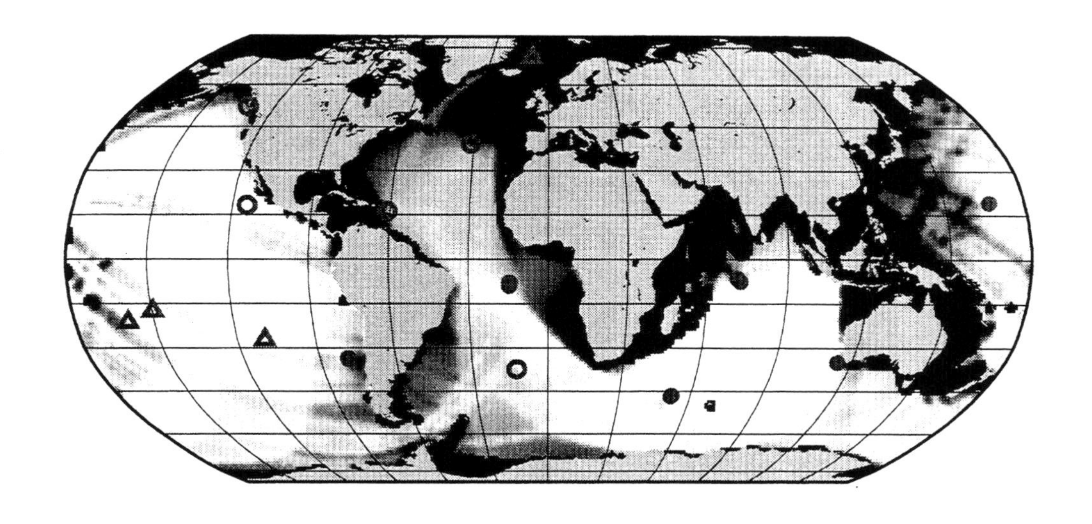

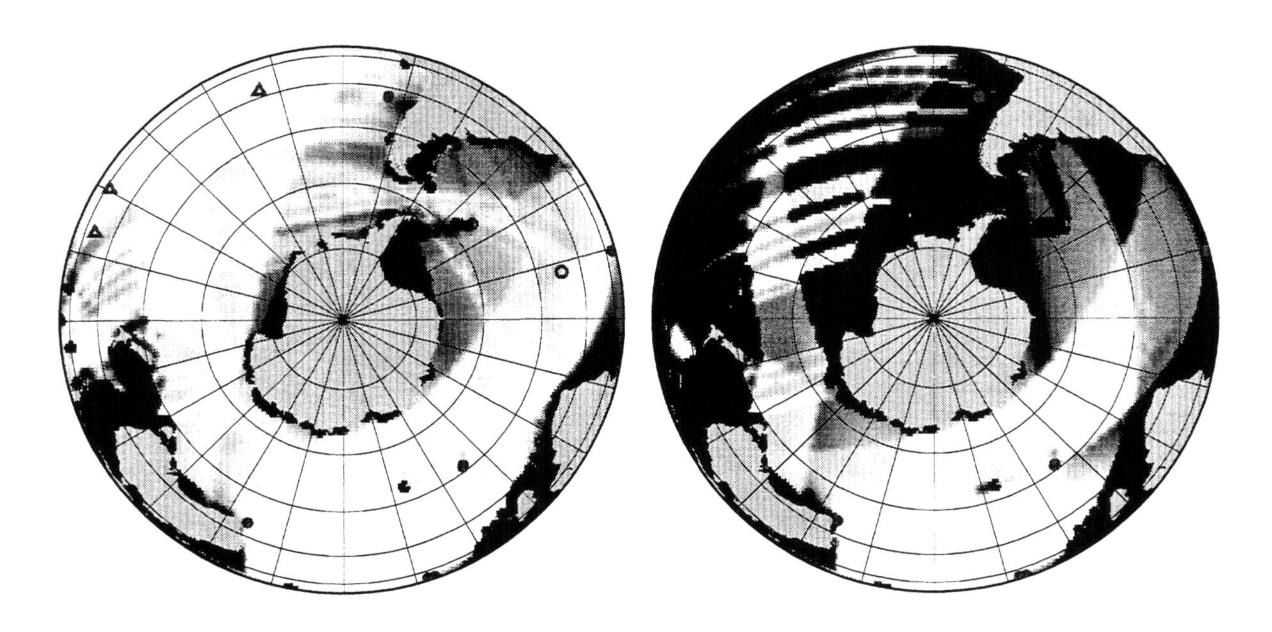
0
1000
2000
3000
4000
5000
6000

be upwardly refracted and heavily scattered by the sea surface and seafloor. Further, sound generated in the Antarctic must pass through a highly variable region of circulating currents (Antarctic Circumpolar Current), which will refract the energy in unpredictable directions, thereby reducing the precision in both detection and localization.

Comprehensive studies of the effects of bathymetric and continental blockage and the propagation characteristics of shoaling waters must be conducted. From a technical viewpoint, the use of multiple hydrophones at hydroacoustic stations, rather than the recommended single sensor, should be considered. The incremental cost is small; the gain can be large. Furthermore, the false-alarm rate of a sparse network of single hydrophones is likely to be prohibitive. The use of two- or three-hydrophone systems that can provide an estimated azimuth, for example, will reduce this problem substantially.

The level of performance to be expected from the five island seismic T-phase stations needs to be determined. The use of land seismic stations to detect ocean acoustic energy that couples to the solid Earth and is transformed into elastic waves is poorly studied and little understood. For this reason, the T-phase stations' contribution to the location accuracy in Figure 3.4 is uncertain. Figure 3.4b shows that in certain regions, the T-phase stations are assumed to provide the majority of the locational capability. For these reasons, it will be important to calibrate existing sensors using artificial sources and to ensure that new installations are well calibrated.

Perhaps not a major research issue, but a general technical issue that must be addressed, is the confusion that arises as a result of nonstandard nomenclature and sound reference level standards. Because the study of T-phase signals (those that have coupled into waterborne acoustic modes and vice versa) by the seismic community is not as mature as the study of solid Earth phases, it suffers from a lack of standard nomenclature consistent with accepted seismological conventions. The effect is to confuse discussion, inhibit communication, and ultimately forestall research progress. Similarly, instrumentation and measurement standards are required to facilitate comparisons of projected performance levels and the results of observations made at widely dispersed locations. At present, for example, the hydroacoustic monitoring community refers all signal levels to 1 microvolt at the receiving hydrophone. The rest of the hydroacoustic community refers signal levels to a pressure level of 1 μPa at 1 m.

Detection

The ocean and ocean bottom are as variable as the atmosphere and internal structure of the Earth. Consequently there are regions of good energy propagation and regions where coupling of source energy into acoustic energy is relatively inefficient. In the former, detection may be relatively straightforward; in the latter it may be impossible.

Large morphological features on the seafloor present particular difficulties for long-range detection of sources of interest. Large oceanic plateaus and chains of islands are likely to block the propagation of ocean acoustic waves. Even relatively small features such as abundant seamounts may block propagation to receivers at the frequencies typically used for hydroacoustic detection of earthquakes and explosions. A thorough search for "blind spots" should be conducted, and two- and three-dimensional propagation codes can be used to explore the effects of seafloor structures. Numerical simulations should be backed up with experimental verification, which might include the use of earthquake sources as well as seagoing expeditions to explore relevant issues. Most of the knowledge of long-range acoustic propagation is restricted to deep oceans. In such an environment, signals from very small sources in the sound channel can propagate thousands of kilometers. However, if the source is located on a continental margin or in shallow water in general, long-distance propagation of acoustic energy is seriously impeded through both surface and bottom interactions, and coupling into the sound channel for long-range propagation can be problematic. It is important to examine these effects so as to develop an understanding of shallow water source region effects on long-range detection. Trade-offs with the generation of seismic waves should be an integral part of such studies.

Although algorithms exist to test the effects of seamounts, oceanic plateaus, and continental margins on the propagation of acoustic energy, it is essential that predictions be tested effectively against observations. Toward that end, scientists should exploit natural sources such as earthquakes and volcanoes for experimental verification. Some seagoing experiments with artificial sources may also be necessary.

Although it is common to observe underwater acoustic energy from earthquakes, volcanoes, and exploration sources at great distances, the coupling (or conversely the blockage) is poorly understood. Sound channel or waveguide propagation in the oceans is characterized by a very narrow band of phase velocities which can be measured in tens of kilometers/second. Most of the common sources (e.g. earthquakes, shallow ships) do not generate such phase velocities in a direct manner. In these cases, propagation depends strongly on heterogeneity and scattering for excitation. This phenomenon is not well understood. For example, an earthquake at a depth of 300 km cannot introduce energy directly into the channel but must rely on the interactions of elastic and acoustic waves at the seafloor to introduce sound into the channel. To date, the mechanism for this transformation has been studied only at the most rudimentary level. It is important that this aspect of long-range propagation be explored in order to understand the signals and noise observed in the water column. For example, there is no correlation between seismic magnitude and amplitudes observed on acoustic sensors located in the sound channel.

Problems of nuclear event monitoring in the ocean are exacerbated by the background of several thousand suboceanic earthquakes exceeding magnitude 4.0 per year. Furthermore, the slow acoustic propagation speed, coupled with the small number of sensors at large separations, enhances the opportunities for false alarms and incorrect locations for small events. Only through the development of robust automatic detection and noise rejection algorithms, coupled with synergy with other monitoring technologies, can the proposed IMS hydroacoustic system be useful.

Association and Location

The hydroacoustic network will rely on the association of arrivals from at least three different stations to perform cross-fixing and location of the event. This is called intra-hydroacoustic association, and it must be done successfully if the hydroacoustic network is to have utility as a stand-alone system for locating hydroacoustically coupled events, especially events that may only be well coupled in the water volume. Two-station location is, of course, possible if the left-right ambiguity can be resolved due to blockages, the positions of land masses, or contributions from other data types (e.g., seismology). Hydroacoustic observations have never been integrated with seismic and infrasound data on the scale proposed for the IMS. It is expected that there will be significant value in combining and associating hydroacoustic, seismic, and infrasound phases. Successful data fusion is expected to have a significant impact on the effectiveness of the hydroacoustic and T-phase detectors because of their small number and sparse distribution.

Several questions and research issues related to hydroacoustic location are raised:

- What level of calibration is required to provide sufficient location accuracy to identify regions for debris collection? Will there be a need for more hydroacoustic stations? Preliminary analysis has shown that when only hydroacoustic data are considered, false alarms create many false associations that lead to inconclusive or erroneous locations, and this problem is greatly exacerbated by the small number of stations and the slow speed at which signals propagate in the ocean. One alternative that should be investigated is adding one or two more hydrophones to IMS stations, with a few kilometers' separation. This will permit determination of a direction or bearing that can be used to improve the intrahydroacoustic association, localization, and false-alarm rejection without achieving a submarine tracking capability.
- Can a single station localize an event by exploiting lateral multipaths due to ocean refraction and/or reflections from bathymetric features such as islands and coastlines, thereby partly mitigating false alarms and improving the location capability of the sparse IMS hydroacoustic array? Understanding diffraction effects and wave number refraction by shoaling bathymetry may also be important for this effort.
- Is the present knowledge of ocean climatology sufficient to achieve 1000 km^2 location accuracy worldwide? It is believed that in large portions of the central open oceans, climatology is not a limiting factor, but in regions of high variability (e.g., major boundary currents such as the

Antarctic Circumpolar Current), this is almost certainly not true.

- What is the performance of T-phase stations for detecting hydroacoustic signals and their consequent contribution to the location capability of the hydrophone network?
- The IMS essentially locates events in latitude and longitude, with the initial assumption that they are at the surface. A significant research issue is to ascertain what can be done to determine event location in depth or at altitude as well—in the ocean, beneath the seafloor, or above the ocean surface—which can be an important basis for identification. In addition to research on using the characteristics (spatial and temporal) of hydroacoustic signals alone, combining hydroacoustic data with seismic data (for underwater and subocean bottom events) and infrasound data (for atmospheric events over the ocean) needs to be investigated.

A further research topic is the calibration of source coupling, propagation paths, and losses for specific hydroacoustic sensors using events that are identified and located by seismic sensors. Improved understanding of the coupling losses and direction changes that occur as energy is coupled from the solid Earth to water and vice versa represents a major research area that must be pursued to use T-phase data effectively at hydroacoustic stations.

Identification

In contrast with chemical and nuclear explosions on land, the data involving explosions (especially less than 1-2 kt) at sea or at low altitudes above the sea surface is limited. Thus, the discriminants presently thought to be useful for distinguishing explosions from earthquakes, volcanoes, and other events at sea have not been subjected to statistically significant testing. Such testing with nuclear explosions will obviously be impossible in the future. Therefore, questions about the efficiency and robustness of present discriminants or the performance level of proposed discriminants will be difficult to evaluate, and alternative means of testing these important algorithms must be developed. At issue is whether nonnuclear-scaled testing is valid.

A similar problem is that many of the methodologies for computing explosion yield and the effective source pressure (in the linear range of propagation) rely on well-known relationships for chemical explosions. Given that the physical processes associated with nuclear explosions produce some different effects than those for chemical explosions, is this valid? Unfortunately, there are no good recordings of previous waterborne nuclear explosions, which require that this relationship be validated with analytic or numerical techniques.

Additionally, the signature of a low-level atmospheric explosion detected by hydroacoustic sensors is not known. Existing evidence indicates that it is composed of relatively low frequencies and that the signal duration is short. There are few data and experiments; calculation and modeling are required.

The high-frequency content of T-phases from earthquakes appears to decay more rapidly than for explosions. It is not known if this relationship holds for small events, nor are the physical reasons for this differential decay known.

Finally, although hydroacoustic and T-phase stations cover a fairly large portion of the world where seismic stations are sparse, their ultimate performance must be understood in the context of the larger system, and the improvements to be realized by the addition of infrasonic and seismic information must be evaluated.

Summary of Research Priorities Associated with Hydroacoustic Monitoring

In summary, a prioritized list of research topics in hydroacoustics that would enhance CTBT monitoring capabilities includes:

1) Improvements in source excitation theory for diverse ocean environments, particularly for earthquakes and for acoustic sources in shallow coastal waters and low altitude environments.
2) Understanding the regional variability of hydroacoustic wave propagation in oceans and coastal waters and the capability of the IMS hydroacoustic system to detect these signals.
3) Improved characterization of the acoustic background in diverse ocean environments.

4) Improving the ability to use the sparse IMS network for event detection, location, and identification and developing algorithms for automated operation.

3.3 INFRASONICS

Major Technical Issues

Atmospheric infrasonic detection in some respects has less complex propagation challenges than hydroacoustic or seismic detection because the medium does not contain discontinuities, such as islands, located in the propagation path. Even the heights of mountains are small compared to atmospheric propagation heights over long distances. However, the propagation of infrasound is affected strongly by rough surface scattering (scatterers are likely to be power-law distributed over a fairly broad range of wavenumbers) and by the presence of winds.

Two additional factors affect the propagation of acoustic waves in the atmosphere:

1) Winds and temperature variations affect their velocity, although there is little attenuation at low frequencies. Typical atmospheric winds can have Mach numbers of 0.1 to 0.2 (about 30 to 60 m/s), which means that the effects of refraction and changes in wind-induced propagation time are important. Also, temperature changes significantly with height. In complex ways this contributes to the creation of a variety of waveguides and ray paths. Furthermore, both horizontal as well as vertical gradients of wind speed and temperature can have important effects.
2) The atmosphere is spatially complex (over scales from meters to hundreds of kilometers) and temporally dynamic (over scales from minutes to months). For slowly varying features, climatological calibrations can account for the effects of long-term trends. For rapid variations, there is a need to develop short-term statistics. Bush et al. (1989) provide evidence for the short-term variability of infrasonic propagation paths. Technical aspects of sound propagation in the atmosphere are discussed in Appendix F.

The dynamics of the atmosphere have an even greater effect on radionuclide transport since the same waves, eddies, shears, temperature gradients, and turbulence that complicate infrasonic propagation, act over longer time periods and cause parcels of air to take complex paths that are difficult to predict. Moreover, the atmosphere is typically not laminar and the turbulent transport fluxes of mass and momentum can be a factor of 10^5 greater than those of a fluid, which might be described with a smooth molecular viscosity-controlled flow.

Some basic operational issues arise for global infrasonic monitoring, characteristic of the research issues that dominated seismic monitoring in the early 1960s when global seismic networks were first being deployed for treaty monitoring purposes. However, extensive experience gained in the 1950s and 1960s indicates that infrasonic monitoring may prove to have relatively few operational limitations for CTBT monitoring. Appendix F provides background and further perspectives of technical issues in this area. From a U.S. national perspective, an assessment of research needs should consider that infrasonic monitoring of atmospheric explosions over broad ocean areas complements satellite coverage (except for cases of heavy cloud cover). Infrasound will provide synergy with seismic methods by identifying the presence of surface effects in chemical explosions and lonesome Lg waves, as discussed in Section 3.1.

Detection

The history of automatic infrasonic monitoring is limited. Consequently, a high priority will be to gain experience with the performance of automated systems for detecting continuous infrasonic signals. Many of the research areas discussed below are thus at a more basic technical level than the other monitoring technologies. Phase identification for atmospheric signals is also poorly understood, and reliable techniques must be developed to make detection useful. A wide range of basic problems in sensor performance, array design, sampling rate, and signal processing is associated with the detection capabilities of the infrasonic network. Given these constraints, the following subsections describe important research activities required to enhance the detection capabilities of the proposed IMS infrasound network.

Development of a Sensor with Extended Response, Temperature Stability, and High Accuracy

The development of a sensor with the required IMS operational characteristics should be a high priority. A critical goal of this work will be to reduce the noise level of the sensor from electronic and other sources (e.g., temperature-induced noise) to at least an order of magnitude below typical minimum atmospheric pressure signal levels (> 0.01 Pa). This means that the sensor noise level should be smaller than 1 mPa from peak to peak over the frequency range of interest. This goal is achievable.

Other sources of infrasonic noise (not associated with the sensor) are pressure fluctuations in the turbulent atmospheric boundary layer and the background of infrasound arising from geophysical sources. With the exceptions of quarry blasts, other chemical explosions, and missile launches, relatively few human processes produce sounds that could mask nuclear explosions.

Past infrasonic sensors operated effectively for years, but they were large, and heavy and did not incorporate current technology. Improvements in thermal insulation techniques should enable the size and weight to be reduced while retaining excellent temperature stability.

Improved Spatial Filter Design

Until the late 1970s, noise from pressure fluctuations in the boundary layer was reduced by using a 300 m long pipe with a sensor at the midpoint. This pipe was tapered in diameter in sections, with small pipes (e.g., 1 cm diameter) at the ends and larger pipes (e.g., 5 cm diameter) at the center. Each pipe was equipped with flow resistors (330 acoustic ohms) at intervals of 2 to 3 m to couple the atmospheric pressure. These installations were difficult to construct and maintain since 200 flow ports had to be kept clean and in the correct range of flow resistance. These noise filters had dimensions smaller than the wavelengths of the nuclear signal sounds they were designed to detect (e.g., acoustic wavelengths greater than 1 km) and dimensions larger than the scales of turbulent eddies producing pressure noise. Thus, they averaged pressure fluctuations from the eddies and did not affect the acoustic waves, which appeared coherently over the spatial filter (producing improvements in the signal-to-noise ratio of typically 20 dB).

Significant design changes in the 1980s, based on the work of Daniels (1959), Bedard (1977), and Grover (1971) allowed detection of infrasonic frequencies near 1 Hz. These designs involved the use of porous irrigation garden hose in place of the pipes with flow resistors. Hoses are continuously sensitive to pressure fluctuations along their entire length and provide additional averaging. These were deployed in a configuration with 12 hoses (typically of 8 or 16 m lengths) radiating outward from the sensor.

There is a continuing need to optimize the design for reducing the noise at lower frequencies while ensuring that signals-of-interest are not affected. This is a critical area of research and operational need that would help to lower the thresholds of signal detection. The panel emphasizes that research on the efficacy of different spatial filters should always involve comparisons between natural signals and noise pressure fields using the same type of microphone for each of the spatial filter designs.

Improved Knowledge of Noise Sources in the Turbulent Boundary Layer

Knowledge of the mechanisms that produce changes in pressure in the surface boundary layer must be improved. Such knowledge can guide site selection and the optimization of spatial filter designs (currently more of an art than a science). Few data sets exist on the scales of pressure fluctuations, and these tend to be episodic and limited in scope. A recent study (Bedard et al., 1992) found that depending on local conditions, the root-mean-square (RMS) pressure noise dependence on the mean wind speed ranges from wind speed to the first power to wind speed to the third power.

Other uncertainties exist. What are the relative roles of temperature and velocity fluctuations in creating pressure noise? Can the boundary layer be modified to reduce noise? These and other questions pose important analytical and experimental research problems. The development of an omnidirectional, all-weather, static pressure probe (Nishiyama and Bedard, 1991) would permit a range of experiments not previously possible. The research payoffs again could lower detection thresholds significantly.

Improved Array Design

For infrasonic signal detection there is evidence that arrays of sensors are required to separate acoustic signals from local wind noise (often, even after reduction by spatial filters, these can have similar amplitudes). The local noise is not correlated from array sensor to sensor for a well-designed array. Arrays of individual sensors also are required for identifying distant sources since locally determined azimuth information is needed to distinguish among multiple sources. The challenge is to optimize the number of sensors and the geometry of their deployment for detection and discrimination.

Once a sparse array design has been defined it is important to explore the pros and cons of adding additional elements and to provide guidance in this regard. Considerations include

- providing additional choices of combinations of sensor sites to use in cross-correlation, thereby increasing the probability of reducing local noise;
- reducing side lobes through choice of number of elements and geometry;
- increasing azimuth and phase speed resolution;
- possibly enhancing the array with directivity options; and
- providing means for discriminating between signal types based on spatial decorrelation as described in the next section (e.g., Mack and Flinn, 1971).

Using filled arrays of instrumentation with improved spatial filters, one can examine the correlation of the infrasonic signal across the array. In some cases, this may be an important discriminant. For example, infrasonic signals generated by ocean waves show rapid decorrelation with distance across an array (these "microbaroms" are the atmospheric analogue of microseisms, both being excited by nonlinear interaction of opposing swells in the oceans) (Donn et al., 1967). These sounds occur almost continuously during the winter months and have the potential to mask sounds from other sources (Posmentier, 1967). Because they originate from large areas of incoherent sources (either interacting oceanic gravity waves or waves impacting beaches), they decorrelate across an array much more rapidly than point sources of sound. Although microbaroms typically have periods from 3 to 6 seconds (wavelengths of 1 to 2 km) they are almost completely decorrelated over array sizes of 4 km and will surely complicate detection for arrays much less than this size. Point sources remain correlated over arrays of 4 km size and larger.

High-Frequency Sampling of Infrasound Signals

High-frequency sampling of atmospheric infrasound, between 1 and 5 Hz, provides potentially important discrimination information that will be valuable for synthesis with seismic data sets. The use of these higher frequencies is an area of continuing research. Examples of signals having significant acoustic power above 1 Hz are described below. Further details for these and other signals are presented in Table 3.1.

- Avalanches. The infrasonic signatures for avalanches are quite unique and usually consist of a wavetrain of duration less than 1 minute with a sharp frequency peak between 1 and 5 Hz. Large avalanches produce detectable seismic energy, and such signals could be identified by data fusion with infrasonics.
- Quarry blasting. Quarry blasts are detected frequently. Their acoustic signatures contain higher-frequency components that can distinguish them from underground nuclear explosions.
- Severe weather. Growing knowledge of the spectral content of infrasonic signals from severe weather suggests that there are important discriminants for identifying and distinguishing these signals from other sources. Current results concerning the relationships between infrasonic signals and storm dynamics indicate that frequencies to at least 5 Hz are important.
- Meteors. The infrasonic signatures associated with meteors can often be distinguished from

TABLE 3.1. Summary of the Properties of Infrasound Generated by Several Different Source Mechanisms (Green and Howard, 1975).

WAVE TYPE →	EARTHQUAKES		VOLCANO and METEOR IMPACT	AURORAL	MICROBAROMS	FRONTAL PASSAGE	SEVERE WEATHER	JET STREAM	MOUNTAIN ASSOCIATED	BOUNDARY-LAYER GRAVITY WAVES
NATURE OF SOURCE →	IMPULSIVE via Seismic	IMPULSIVE Direct Air Wave	EXPLOSIVE	SERIES OF IMPULSES (Bow Shock)	CONTINUOUS (Air-Sea-Wave interaction)	PRESSURE JUMP & BUOYANCY OSCILLATIONS	UNKNOWN (1. Penetrative Convection 2. Lightning 3. Turbulence)	UNKNOWN (Shear Instabilities)	OROGRAPHIC	AERODYNAMIC: SHEAR INSTABILITIES
OBSERVED CHARACTERISTICS ↓										
WAVE PERIOD (Seconds)	12-25 (0.5-80)*	12-25 (1-30)	1-300 (0.5-1500)	10-400	5-7 (2-8)	180-1500 (180-1800)	12-60 (6-300)	(240-7200)	20-70 (10-120)	300-900 (250-1500)
WAVE AMPLITUDE (μbar O-peak)	1-10 (0.1-20)	1-2 (0.1-20)	up to 5000	1-5 (0.2-20)	0.1-1 (0.1-5)	(50-2000)	0.5-3 (0.1-20)	50-200 (10-1000)	0.5-3 (0.1-10)	25-200 (10-1000)
DURATION	up to 1 hr. Depends mainly on distance and dispersive properties of media.			Seconds to hours	Hours to Days (Noise Limited)	Minutes to Hours	Minutes to Hours	Hours to Days	Hours to Days	Up to 1 Hr.
HORIZONTAL TRACE SPEED (m/sec)	3k-15k	340 (320-360)	320-500	400-1200	340	2-20	330-360 (330-450)	15-75	340 (330-360)	5-20
DIRECTION OF TRAVEL	Vertical	from Epicenter ±6°	from Source	from intersection of Midnight Meridian with Auroral Zone	from certain Oceanic "Centers of Activity," e.g., North Atlantic Gulf of Alaska	Direction of Frontal Travel	from Centers of Strong Convective Activity; Mainly Midwest	Flow Direction	from certain Coastal/Mountain locations; e.g., West Coast of Canada, Argentina/Chile.	Mean Wind
TRANSVERSE COHERENCE DISTANCE (to zero correlation)	not measured		~ 14 λ	Large at Low Latitudes; 10° Azimuth & Elevation Fluctuations	~ 2 λ	not measured	~ 14 λ	not measured	~ 14 λ	1.15 λ
DISTINGUISHING PROPERTIES	Seismic Dispersion	Long Periods precede Short Periods		High Speeds; Short Impulse precedes Long Periods; Observed at Magnetic Latitudes > 30°	Nearly Monochromatic; Beaded Wave Envelope; Low Coherence	Large Amplitude; Often Wavelike	Azimuth Change with Time; Spring / Summer Peak	Low Phase Velocities	Fixed Azimuth for up to Days; Local Winter Occurrence	Probably Exist down to Smaller Scales than Presently Studied; Usually Advected by Mean Wind.

*Figures indicate nominal values; those in parentheses indicate extremes.

explosions based on the frequency content. Recently, a bolide was tracked over the southwestern United States using the signal above 1 Hz (ReVelle, 1995).

- Mountain-associated waves. Infrasound signals are generated by airflow over mountain ranges. Acoustic waves with periods between 50 seconds and 1 second often are excited. Energy greater than 1 Hz also occurs. The existence and character of high-frequency energy may be valuable for identifying this class of signal, which can be detected at distances of thousands of kilometers (Larson et al., 1971; Green and Howard, 1975).
- Volcanoes. This is another area in which valuable discrimination information may exist at higher frequency. It should be an area for future research (Buckingham and Garces, 1996).
- Aurora. For IMS stations at high geomagnetic latitudes impulsive infrasonic signals of amplitudes up to ten microbars with periods as short as ten seconds are frequently observed, on the night side of the earth, as bow waves generated by supersonic motions of large scale auroral arcs that contain strong electrojet currents, (Wilson, 1971). The specific source arcs that generate these impulsive type auroral infrasonic waves (AIW) can not be identified by triangulation, using the successive arrival of the AIW waves at two stations, because of the highly anisotropic nature of the bow wave radiation by auroras (Wilson, 1969a). However, if the auroral source arc of an AIW passes over the infrasonic station there will be a large magnetic perturbation that is easily observable and can be used to identify the infrasonic wave as one of auroral origin. Past observations in polar regions, especially by the University of Alaska, provided much useful data on these phenomena.
- Earthquakes. Colocated seismic and acoustic arrays offer opportunities to identify seismic energy coupled from the atmosphere at frequencies greater than 1 Hz (as well as lower frequencies). Conversely, it will also be valuable to identify infrasonic signals caused by local seismic disturbances (Donn and Posmentier, 1964).

The optimum infrasonic sample rate should be at least 20 Hz. This will ensure that the acoustic signals are completely defined and that no information important for discrimination or yield estimation is lost. It should be done at least during the early stages of the IMS until it can be determined if a lower rate can be employed. In addition, a growing number of geophysical monitoring applications are being identified that represent potential valuable resources for other national needs (e.g., avalanche and tornado detection and warning, as well as monitoring turbulence aloft). These applications are dependent on data up to 5 Hz.

Improved Signal Processing

Array processing is required for typical infrasonic signals because signals with large signal-to-noise ratios tend to be the exception, except for nearby or large geophysical events such as volcanic eruptions. Even these events at long ranges can be comparable in amplitude to background signal threshold levels. On the other hand, the fact that wind-induced pressure noise will be uncorrelated from array element to array element offers a means to reduce this class of contamination. An analog cross-correlation was used prior to the 1970s to perform processing, and a digital version of this instrument was produced by Young and Hoyle (1977). This technique has been applied to the processing of atmospheric gravity waves (Einaudi et al., 1989) and ocean waves.

The use of a data-adaptive Pure-State Filter (PSF) (Samson and Olson, 1981; Olson, 1983) is effective in reducing incoherent, isotropic wind noise in multivariate infrasonic array data. Non-isotropic, incoherent noise can also be eliminated from infrasonic array data by adapting the PSF filter characteristics to the data in regions free of signals of interest (see Olson, 1982). While a wide variety of analysis techniques have been used in the search for infrasonic signals in the data stream, such as beam-steering, f-k analysis, and cross-correlation combined with least-squares estimators, the performance of each technique is always improved when the data are PSF filtered prior to analysis (Olson et al, 1982). An increase in signal-to-noise ratio of as much as 20 db and an increase in the maximum value of inter-microphone cross-correlation coefficient by as much as 0.20 have both been obtained after PSF filtering of the raw data. Further research on the application of data-adaptive, frequency domain filters to infrasonic data should be pursued to enhance the detection of low level CTBT monitoring station infrasonic signals.

There are a number of methods for improving the ability to distinguish signals and to visualize results. A range of geophysical phenomena can produce false alarms, mask nuclear detonation signals, or complicate analysis. Some of these produce unique signatures that can be identified using computer algorithms. For example, avalanches contain a short-duration, fixed-frequency wavetrain (see Appendix F). One approach to improve signal processing would include the following steps: (1) produce a training set using acoustic signals from known avalanches, (2) train a neural network, and (3) perform testing to validate the network skill in identifying avalanche signals. Such research could then be expanded to all types of signals using a variety of approaches.

Calibration and Testing

Important calibration and testing activities include the following.

1) The IMS will include sensors of a variety of designs, all intended to have the same properties in terms of operating characteristics. Because these designs will vary from country to country, methods of calibration must ensure that sensor sensitivities and frequency responses are within agreed-upon limits. Because few standards for calibrating infrasonic sensors exist, there is a need for transportable standards to perform cross-calibration between countries as well as monitoring sites.
2) Because of the nature of their design and the critical mission they perform, spatial noise reducing filters must be verified and calibrated periodically. A break in the filter hose is the equivalent of an electrical short to ground, and it is necessary to locate and fix such problems. This area of research and development will have a great impact on day-to-day operations.
3) There is a need to calibrate the array function itself at single or multiple sites. Research on controlled infrasonic sources could allow testing of array functions for detection and bearing estimation. They would also provide an opportunity to evaluate array performance. Dedicated large conventional explosive tests are good candidates, but these are not possible in all areas.

Association and Location

The process of association and location for the infrasound network can benefit from the known approaches used in seismology, although some modifications may be necessary. Research in several areas is needed because global infrasound networks have not been operated for more than 20 years. The characteristic parameters of events of interest such as spectral content, duration, amplitude, and correlation require elaboration. Some of these are known from past experience, but incorporation into automated processing has not been performed for infrasound data. Effects of atmospheric conditions on travel times, frequency content, duration, and bearing accuracy need to be understood. Propagation models should include the range dependence of atmospheric variables. Additional analyses of past infrasound data on nuclear explosions as a function of yield and path would be useful.

Size Estimation

Past monitoring experience provides a baseline amplitude-distance relation. Improvements will come from the incorporation of atmospheric dynamics to account for observed signal amplitudes and, therefore, size estimates. Fresh examination of the relationship between known period and yield for low-yield events would contribute to improved estimates. Additional research on atmospheric signals from partially coupled events will establish a modified amplitude-distance relation as a function of coupling efficiency. However, since yield estimation is of secondary importance to CTBT verification, research in this area is not a high priority.

Source Identification

Infrasonic data taken at long range from an event will not differentiate a chemical explosion from a nuclear explosion. Viewed at long range, both are essentially point sources of impulsive energy release. Thus, there is a need to define unique explosion features. In parallel, the characteristics of natural and human-induced sources must be determined for discrimination purposes. These "background" sources have some known characteristics, but not all sites will see the same background. As sites become operational, their backgrounds will have to be cataloged. This will allow the rapid dismissal of some sources from further analysis.

Some past work and current research have defined signal characteristics of natural and human-induced sources, but much remains to be done. A review of past U.S. and French data from nuclear explosions in the atmosphere may provide a unique explosion "voiceprint." Focused field experiments or extended monitoring to create archival data suitable for source characterization may lead to ways to distinguish nuclear explosion signals from natural or man-made sources. These data bases could also be used to examine the synergy between seismic and infrasonic technologies to study evasion scenarios in which a chemical explosion is used to mask a nuclear explosion. These studies should be coupled with analytical or numerical investigations to understand and predict the spectral power produced and to extend the results to explore the ramifications of source strength or size changes. Once such knowledge bases exist, they can be applied to eliminate such sources as potential false alarms. There should be valuable intersections between the nuclear monitoring and geophysical communities. Some resources that will be worth drawing on include past meteor, volcano, and earthquake detections as well as documented signals from missile launches. A recent compilation from three years of array data at Antarctica and Fairbanks Alaska (Wilson et al., 1996) is an example of a database that will be valuable for these purposes.

Attribution

Locations determined by infrasound are unlikely to be sufficiently accurate to do an immediate sample collection, but they may help guide a wide-area search in the open oceans and focus the processing of signals detected by other technologies. Attribution by infrasound would be possible in cases where the location and error estimates lie totally within a country's borders. In broad ocean areas, attribution will not be possible solely through the use of infrasound data (as is the case for seismology and hydroacoustics).

Summary of Research Priorities Associated with Infrasound Monitoring

In summary, a prioritized list of research topics in infrasonics that would enhance CTBT monitoring capabilities includes:

1) Characterizing the global infrasound background using the new IMS network data.
2) Enhancing the capability to locate events using infrasound data.
3) Improving the design of sensors and arrays to reduce noise.
4) Analyzing signals from historical monitoring efforts.

3.4 RADIONUCLIDES

Radioactivity measurements provide a unique constraint on nuclear testing underground, underwater, and in the atmosphere if the explosion deposits significant amounts of radioactive material in the atmosphere. Nuclear detonations that take place at shallow depths underground, underwater, or in the air near the surface may emit significant amounts of radioactive material into the air. Much of this radioactivity will be on large particulates or droplets that fall out near the site of detonation. Most of the radioactive noble gases formed during the explosion or shortly thereafter will be released and will move away with the local air mass. The radioactivity will remain on or near the surface of the soil, where it can be readily detected for years after the detonation. The radioactivity on or near the water surface will be transported and dispersed by ocean currents, and monitoring will have to be rapid in order to access the location. Appendix G provides a brief review of projected radioactive particulates and noble gases released under a variety of test conditions.

In the time following release, the relative

proportions of radionuclides will be modified by fractionation and rainout effects.[20] The importance of these mechanisms depends on the type release or detonation site (see Appendix G). For example debris from a reactor accident will be highly fractionated because the more volatile fission and activation products are preferentially released. Nuclear device debris will be highly fractionated if the fireball interacts with the surrounding medium. If the medium is soil or rock, there will be extensive localized fallout of the more refractory debris. If the medium is water, there will be localized loss of both refractory debris and the more soluble radioactive gases. Further fractionation takes place during the subsequent transport of the debris and gases by the atmosphere or, in the case of an undersea explosion, by water currents. Substantial fractionation has been observed even in high altitude tests. Because of the lower temperatures involved, low yield devices are usually more fractionated than high yield devices detonated under the same conditions. Rainfall can lead to the fractionation especially of soluble gasses such as isotopes of iodine and bromine. Certain waterborne organisms will also take up particular radionuclides. For example, plankton will concentrate plutonium at levels 2,600 time that of surrounding waters. Because of fractionation effects, it is important to collect and analyze samples as soon as possible after detonation.

Deep underground detonations will result in the containment of almost all of the radioactive fission fragments under ground, with the possible exception of some of the radioactive noble gases. Over time, these gases may escape to the air above the site through cracks and fissures in the rock and soil. Most of these gases are produced by the decay of precursors and will be released hours to weeks after the detonation; their activity will diminish slowly with time. The amount of gas released at different times will also depend on the atmospheric pressure over the site, with increased release under conditions of low pressure and reduced or no release under high-pressure conditions.

The present plan for remote CTBT radionuclide monitoring is to install a series of fixed stations around the world that have fully automated measurement devices with the ability to collect, process, and send the results to a central station on a daily basis. Alternatively, a sample may be sent to the central facility for processing. In both scenarios the data are subsequently evaluated using intelligent algorithms for positive or negative testing of event signals. The technology for these functions is developed, but the technology for the fixed stations has not had extensive field testing to ensure long-term reliability, especially in remote areas. Research can be pursued to develop more rapid and mobile monitoring capabilities to include air, water, and land monitoring. The sooner radionuclide samples can be collected and analyzed after a test, the greater will be the amount of information obtained about the nature of the test device. For instrumental analysis of airborne radioactive particulates collected on filter paper, the lower limit of detection is determined by the amount of naturally occurring and extraneous human-made radioactivity also collected in the sample. If radiochemical separations are employed, the limit of detection depends only on the activities of the separated fission products. The limit of detection of the radioactive noble gases on a separated gas sample is determined only by interference from other radioactive gases in the sample and by the activity of the gas of interest.

The delay time associated with the transportation of the radionuclides to the fixed monitoring locations may be days to weeks. Once a clear radionuclide event is detected, there is a need to backtrack the atmospheric transport history to locate the source of the event. This requires the use of models to propagate the transport and dispersion of radionuclides back in time based on archived measurements of atmospheric properties (wind speeds, temperatures, etc.). Such efforts are limited by the quality and density of available data and fundamental uncertainties on modeling chemical transport in turbulent atmospheric flows. These same models may be used in a predictive mode (e.g., to direct sampling efforts) using the output from forward calculations that predict atmospheric properties and circulation. Because other monitoring technologies will provide rapid

[20] Fractionation is the preferential loss of specific radioisotopes after the initial release. The degree of fractionation increases with time, caused by a combination of physical forces (such as gravity) and chemical processes (such as oxidation). Rainstorms will also "wash" radionuclides out of the atmosphere (i.e., rainout). This is important for all radionuclides except for the noble gases.

data about the time and location of a potential nuclear explosion, forecasts of radionuclide transport could be used to vector aircraft for sampling closer to the source than would be possible with the fixed radionuclide stations. A capability to back-track ocean currents may also be needed for tests in the ocean or near the ocean surface. Basic research is needed to enhance back-tracking capabilities in general, particularly to improve the accuracy of projections beyond 5 days and to provide realistic uncertainty assessments.

For On-Site Inspection, radioactive gases leaking from cracks and fissures may be the only surface indicator of the detonation location. Rapid noble gas monitoring equipment mounted in slow-moving aircraft to perform the initial screening of an area may identify the approximate test location; however, the escaping gas may not be present in sufficient amounts to allow detection. Such aerial surveys may require the collection and rapid analysis of many air samples over a wide area.

The detection and identification capabilities of the network of radionuclide detectors (particulate and gas) will be determined by

1) minimum detectable concentrations of the system for particulate isotopes,
2) spatial and temporal variations in the radionuclide background at individual stations,
3) atmospheric or waterborne transport processes that disperse the fallout and lower the measurable concentrations of isotopes,
4) radioactive decay processes that reduce the concentration of particular isotopes, and
5) fractionation and rainout processes that selectively remove particular isotopes from the air, water, or land.

Considering the above issues, an assessment of detection capabilities involves calculations of radionuclide transport away from the site of a detonation to downstream detectors. This modeling requires assumptions regarding fallout/release from different detonation scenarios, atmospheric and ocean transport processes, detector efficiencies, and fractionation and rainout efficiency. Modeling to date, has largely considered the first three factors, independent of radioactive decay processes. Once radionuclides have been measured in the field, identification requires the ability to distinguish the chemical signatures of nuclear explosions from reactor emissions. Preliminary analysis for atmospheric transport of xenon radioisotopes suggests that this task may be difficult more than 2 weeks after an explosion because of radioactive decay processes. In general, the panel concludes that there is a need for a full assessment of the detection and identification capabilities of the radionuclide network, similar to the modeling that has been carried out for the seismic and hydroacoustic systems (e.g. Figures 3.2, 3.3, and 3.4).

Major Technical Issues

Fixed Station Air Particulate Monitoring

The IMS will include surface stations for collecting air samples for radionuclide analysis. All of these stations will be equipped with some type of particulate collection system that involves the collection of atmospheric dust on filter paper. In the proposed U.S. equipment, radioactive material on the filter paper is then counted by using a high-resolution HPGe (high-purity germanium) detector, and the resulting gamma spectrum is analyzed by using computer-based gamma-ray spectroscopy. As shown in Appendix G, only tests conducted above the transition zone in soil (within 100 m of the surface) or at shallow depths in water are likely to result in a significant release of radioactive particulates. In general, this type of monitor will be of little use in detecting events that are contained or are not vented.

This particulate detection technology is mature and reliable. The panel concludes that the automated system developed by DOE seems to be designed around the best existing technology. However, operational problems associated with the automated daily collection of aerosols on filters, sample counting, and transmission and storage of data may be challenging. Also the amount of spurious radioactivity collected in the dust can produce background radiation that inhibits good signal-to-noise ratios, thus limiting the sensitivity of gamma-ray counting on a daily basis. A clear plan for designing and implementing a gamma-ray system for daily use must be spelled out in great detail and tested for detection limit capabilities. The use of low-level counting techniques (e.g., shielding, Compton suppression, or other coincidence methods similar to xenon identification) may also have to be investigated to achieve the best sensitivities available.

A major problem for fixed station radionuclide monitoring is the establishment and maintenance of a reliable monitoring network in remote locations. Many of the proposed 80 radionuclide sampling sites will be in remote locations without dependable power supplies and technical support. Existing stations are located mainly in developed countries and at sites that have reliable technical support and available backup monitoring systems. Particulate monitors all employ high-resolution gamma-ray spectroscopy using large-volume HPGe detectors that must be cooled continually to a low temperature with either liquid nitrogen or electric coolers. Should a detector warm up to ambient temperature through the loss of power or liquid nitrogen supply, it would take many hours before the detector could be brought back on-line and days should the detector's electronics become damaged during the warm-up.

High-Resolution Gamma Detector for Ambient Temperature Operation

A research effort is needed to develop large-volume, high-efficiency, dependable semiconductor detector materials that can be operated at room temperature and are insensitive to routine atmospheric temperature changes. This type of detector material would have a larger band gap than HPGe, which would allow it to operate at room temperature. Because of the larger band gap, the resolution of detectors using this material will be slightly poorer than the current HPGe detectors but far superior to NaI(TI) detectors. Work on a variety of proposed detectors of this type has been under way for many years. The difficulty in development is incomplete collection of all of the radiation-produced charge pairs. Trapping, especially of the positively charged holes, distorts the resulting gamma spectrum. Small detectors of this type have been developed and are available commercially; they include detectors made from cadmium telluride, cadmium zinc telluride, and mercuric iodide. Research is also under way to make detectors from gallium arsenide, lead iodide, and other semiconductor materials having band gaps in the range of 1 to 2 eV (electron volts). Successful development of this type of detector would provide a rugged, portable, and easy-to-use gamma-radiation monitor that could be started quickly and battery operated. This would be ideal for use in fixed stations in remote areas and for mobile monitoring applications.

Fixed Station Radioactive Noble Gas Monitoring

It is a CTBT requirement that 40 IMS stations employ some type of radioactive noble gas monitor. As shown in Appendix G, these gases are more likely to be released than radioactive particulates from deep underwater or underground tests. Also, these gases will tend to separate from particulate matter and will not be lost by particulate fallout or rain-out. If rain-out or other atmospheric conditions remove particulates from the air, the automated xenon analysis system would become the most sensitive detector of fission products. Because of the low air sampling rates of current noble gas monitors compared to particulate monitors, CTBT noble gas monitoring is considered to be about 1000 times less sensitive than particulate monitoring, and there have been questions about how extensively such monitors should be employed in the proposed worldwide network (Ad Hoc Committee on a Nuclear Test Ban, 1995).

The prototype technology being developed by Battelle (Pacific Northwest National Laboratory [PNNL] Automatic Radioxenon Analyzer) collects atmospheric air at a rate of 7 m^3 per hour (Bowyer et al., 1996; Perkins and Casey, 1996). Xenon gas is separated from the other gases by a series of adsorption and deadsorption steps. The xenon gas, containing any radioactive xenon fission isotopes, is then counted by use of coincidence methods (using proven sophisticated electronics) to assay for signals of radioactive xenon. Tests conducted for nuclear power plants have proven that this instrumentation is highly reliable. There are concerns about the NaI(TI) detectors proposed for this system. They will have limited resolution, making it difficult to distinguish the different radioisotopes of xenon because of overlapping spectral lines, especially in the low energy region. Furthermore, the system gain of NaI(TI) detectors is extremely sensitive to ambient temperature since small temperature changes can make significant changes in the system's energy calibration. Thus, these detectors either must use some type of sophisticated spectrum stabilizer or must be operated at a near constant room temperature. It may be difficult to achieve this type of temperature control consistently at remote site locations. Here again, it would be useful to have available a high-resolution semiconductor detector able to operate with stability at room temperature.

A general problem associated with radionuclide monitoring at a fixed site is the waiting time required for the radioactive gas to reach the stations where it can be collected, counted, and analyzed. However, if radioactive xenon is detected, it is powerful evidence that a nuclear explosion has taken place.

The instrumentation for xenon sampling must be field-tested to evaluate its reliability for long periods of time. In addition, identification software must undergo quality assurance and quality control to ensure that the final results do indeed represent an event. The monitoring for radioactive xenon emitted from nuclear power plants can be used for this purpose.

Rapid Response Airborne Monitoring

The IMS will have no airborne radionuclide monitoring capabilities. In the past, airborne detection played an important role in NTM for nuclear test monitoring, and the United States maintained rapid response aircraft equipped with radiation detection equipment that could locate and follow a radioactive plume once a source had been located and identified. The IMS system is focused on a worldwide network of fixed stations, all of which will monitor radioactive particulates and some of which will monitor radioactive noble gases. Although the response from a fixed station seismic, hydroacoustic, and infrasonic monitoring network will be rapid, the response of the fixed station radionuclide monitoring network will be relatively slow. Even when fission products are detected at a station, much information about the type and location of the nuclear event will be lost because of nuclear decay. With the use of evasive procedures, especially bursts in deep underground boreholes or deep mines or deeply submerged ocean tests, there will be few or no radioactive particulates, and potentially little noble gas will be emitted into the environment to be detected by the fixed station monitors. However, if the approximate location of an event identified as a nuclear explosion is determined by seismic, infrasonic, and/or hydroacoustic methods, rapid response teams using monitoring aircraft and, where politically possible and appropriate, surface or ocean surveillance equipment could be in the field collecting samples. These response teams ideally would be equipped with a rapid radionuclide monitoring capability that is mobile and sensitive and can quickly produce quantitative results. According to available information, no such monitoring systems currently exist or are under development.

Of special importance to airborne monitoring is a rapid response noble gas monitor. Noble gases collected soon after a detonation can contain more xenon radioisotopes as well as argon and krypton radioisotopes. Thus, more information can be determined about the explosive type. Depending on test depth and weather conditions, these gases may be the only airborne fission products available for detection.

The Ad Hoc Committee on a Nuclear Test Ban of the Conference on Disarmament showed a considerable interest in airborne monitoring triggered by other monitoring techniques (Ad Hoc Committee on a Nuclear Test Ban, 1995). Some experts in attendance felt that the primary value of airborne monitoring was for atmospheric explosions over ocean areas or underwater, where fission product evidence can be lost quickly. Monitoring would be directed primarily at remote neutral areas not adequately covered by the network of ground-based stations. They discussed the need for both particulate and noble gas monitors on aircraft that would also be airborne laboratories. This element of the IMS was not pursued because of cost considerations.

A relatively rapid radioactive noble gas monitor was developed in the late 1970s for monitoring radioactive noble gas in and around nuclear power stations (Jabs and Jester, 1976). This approach could be investigated for use as a portable, sensitive, rapid CTBT noble gas monitor. For most radioisotopes of argon, krypton, and xenon of interest to this work, such a system could achieve a lower limit of detection (LLD) sensitivity of less than 1 mBq/m3. This detection sensitivity approaches that of the PNNL Automatic Radioxenon Analyzer, but with a sample collection and analysis time of less than an hour per sample.

The basic concept of this system is the rapid

compression of filtered air into a spherical steel chamber at pressures of 100 to 200 atmospheres. A high-resolution germanium detector is positioned in the center of the chamber. Counting for one-half hour can be used to achieve the LLD given above. By using off-line gamma-ray spectroscopy, a single setup could collect and analyze samples at a rate of one sample per hour or less. Such a system could be operated from an airplane searching for radioactive gases within a few hours of a suspected nuclear event. Crew safety and decontamination factors would have to be considered in any system feasibility studies for other than dedicated airplanes.

Proposed Rapid Particulate and Radioiodine Airborne Monitor

Any rapid response team should have available a capability to collect and analyze quickly airborne particulates containing fission products from a nuclear test. There is another important group of fission products that has not been mentioned in any of the proposed radionuclide monitoring programs: the iodine fission products including 129I (half-life, t½ = 1.6 × 107 years), 130I (t½ = 8 days), 132I (t½ = 2.3 hours), and 133I (t½ = 21 hours). Iodine-132 lasts longer in the environment than suggested by its short half-life because of its precursor ^{132}Te ($t_{1/2}$ = 3.3 days). These biologically significant fission products are frequently in a volatile form that will pass through paper particulate filters and will not be seen by the radioactive xenon monitoring systems.

A well-established procedure for the separate collection of particulates and radioiodines would be the use of a train of stacked filter cartridges. These rugged cartridges allow the passage of a large amount of sampled air in a short time. The most common particulate filter would be a Hepa filter cartridge, whereas the most efficient radioiodine filter would be a silver zeolite cartridge. Such a filter train would have to be installed on the inlet of the compressed air noble gas monitor previously mentioned to prevent the interior of the pressure vessel from becoming contaminated with fission products. These inlet filter cartridges could be used to provide the samples for subsequent gamma-ray spectroscopy. The cartridges could be changed at the completion of the collection of each compressed air sample and counted using an auxiliary HPGe detection system. Because of its long half-life, ^{129}I is difficult to detect by gamma-ray spectroscopy but there are extremely sensitive neutron activation analysis procedures for its detection.

It is recognized that the IMS technologies are fixed, and they were designed explicitly to avoid potential monitoring of reactor releases. Thus, much of this technical discussion may apply only to NTM.

Rapid Response Waterborne Monitoring

Instrumentation and research to support waterborne radionuclide analysis are being carried out by groups at Lawrence Livermore National Laboratory, Sandia National Laboratories, and the Naval Research Laboratory. The primary motivation for these projects is to enhance the capability to monitor radioactive materials from nuclear explosions and nuclear reactors that have been dumped in the ocean. These systems will involve sampling by aircraft, remote underwater stations, and buoys. The remote systems will involve communication by satellite telemetry and most will utilize Na(TI) detectors.

To support these efforts, there appears to be a need for research on deep water collection and concentration techniques for the analysis of dissolved and suspended fission and activation products. The most likely method might be a combination of filtration and mixed bed ion exchange that would avoid chemical fractionation. The dissolved gasses could then be extracted from water samples for subsequent noble gas monitoring. To improve the performance of these systems, there is also a need for a high efficiency, high resolution gamma detector that does require a high capacity cooling system.

Improvements in Basic Data Used to Make Source Term Estimates

Sensible recommendations for additional work on source term estimates are contained in a study by experts from Lawrence Livermore National Laboratory (LLNL) and other National Laboratories (see Appendix G). Given the absence of available data, a few experiments would improve the characterization of those source terms that have the highest uncertainties, and only a modest effort would be required to obtain the needed information. These include high-altitude releases using ^{102}Rh and underwater tests using xenon as a tracer.

In addition, a more complete literature search should be conducted to obtain information useful for improving source term calculations. There should be an aggressive effort to obtain such data from old reports and notebooks at National Laboratories, universities, and other U.S. sources. Such data should also be sought from foreign countries that have conducted nuclear tests, as well as from other countries that have maintained an atmospheric monitoring program. Data collected should be made available to the broader scientific community for both monitoring and scientific purposes, to the extent consistent with national security considerations and terms of acquisition. The information sought should include data on radionuclide measurements and the sizes of radioactive clouds as a function of time. It should also include any data on chemical fractionation and atmospheric partitioning of debris that takes place during the release of radioactive materials and gases under various blast conditions. Any information on the movement of radionuclides after underwater tests would be extremely useful.

Improvement in Air Trajectory Models for Backtracking Calculations

Many atmospheric scientists and engineers employ various back-trajectory models to identify the source terms of various airborne emissions such as sulfur and organic or inorganic constituents. Organizations such as NOAA, the Canadian Atmospheric Environmental Services, EPA, and some U.S. National Laboratories and their European equivalents have successfully used various backward and forward trajectory models (Evans, 1995; Mason and Bohlin, 1995). Usually, such models are reliable for only 5 to 7 days of backward or forward trajectories. While there is general agreement among models that predict large-scale atmospheric flows, different chemical transport models used for forecasting or back-tracking often provide different results, particularly if the initial conditions vary slightly. For this reason, there should be a major research effort to improve the algorithms for back-tracking and prediction of chemical transport in the atmosphere. This effort should include an intercomparison of the models as part of program validation. An ideal test would be determination of the location of various nuclear power stations in the Northern Hemisphere based on their emissions of xenon radioisotopes. The automatic radioxenon analyzer being developed by PNNL could be employed to provide data for these tests. Such an investigation followed by a sensitivity analysis would give a clear indication of the accuracy of these models.

Rapid Radiochemical and Instrumental Techniques for Radionuclide Analysis of Filter Papers

To discriminate fission products resulting from nuclear weapons testing from those produced in nuclear reactors, one can establish a list of CTBT discriminant fission products (Wimer, 1997). The basic requirement is to perform an isotope assay of each of these discriminant isotopes in a mixed fission product sample to 10 per cent within 24 hours. Samples would be particulate filter papers collected from field systems such as DOE particulate samplers or from the proposed rapid response teams. Rapid collection and analysis of these samples is again required because of the short half-lives of some products such as ^{97}Zr (17 hours) and ^{143}Ce (33 hours).

It should also be possible to establish a list of specific fission products that can provide information about the nature of a nuclear weapon that has been detonated. Distribution of the fission products formed will vary with the fissile material employed (i.e., uranium-235 or plutonium-239) and the mean energy of the neutrons causing the fissioning. It is felt that some of the fission products of the lanthanides will be relevant (Wimer, 1997). Of the lanthanides, the fission products of europium (i.e., ^{154}Eu, ^{155}Eu, ^{156}Eu, and ^{157}Eu) may be the most useful. Potentially useful lanthanide fission products include ^{153}Sm and ^{159}Gd. Other fission products that may be useful for characterizing the fissile material include ^{72}Zn, ^{72}Ga ^{103}Ru, and ^{111}Ag.

Aside from instrumental gamma spectroscopy, no rapid procedures currently exist for analysis of selected fission products that are collected on particulate filter papers. A determination of the ratio of the activity of one or more of these europium radionuclides to that of lanthanum-140 (^{140}La) would clearly show the type of nuclear fuel used in a test device. Research is needed to establish either a radiochemical procedure or some type of instrumental method to allow rapid determination of these critical fission products in the field.

It is recommended that this problem be investigated to determine whether or not some existing radiochemical procedure can be modified for this purpose. If not, such a procedure should be developed rapidly. An ideal procedure would be an automated chemical separation that isolates the discriminant elements into sample forms suitable for HPGe counting. Chemical yield information ideally would be generated during these separations.

For certain discriminant fission products, instrumental techniques such as gamma-gamma coincidence counting may be useful to pick them out of the gamma background. Although improved observations of a signal in the presence of high background radiation is possible, detection efficiencies are much lower, and the ultimate count time may extend for several days. Also, the complexity of such instruments cannot be allowed to lower the mean-time-between-failure requirement for unattended operations. Procedures must also be established for rapidly transporting, from fixed stations to a laboratory that can perform this type of analysis, those filter paper samples found to contain recently produced fission products.

Other Areas That Should Be Investigated to Improve Radionuclide Detection Sensitivity

Other nonradiochemical or instrumental procedures for low-level counting include Compton suppression and gamma-gamma coincidence methods. These techniques, although mature, have never been fully employed for CTBT monitoring. The primary advantage of such coincidence methods is the enhanced signal-to-noise ratio, as in the case of the beta-gamma coincidence employed in the radioxenon analyzer being developed by PNNL. It is also well known that the larger the detector volume, the greater is the detection efficiency, especially for higher-energy gamma lines.

Summary of Research Priorities Associated with Radionuclide Monitoring

The top priorities are:

1) Research to improve models for back-tracking and forecasting the air borne transport of radionuclide particulates and gases
2) Research and data survey to improve the understanding of source term data.[21]
3) Understanding of atmospheric rain-out and underground absorption of radionuclides from nuclear explosions.
4) Assessment of the detection capabilities of the IMS radionuclide network.

A secondary priority is:

5) Research on rapid radiochemical analysis of filter papers.

A long term research priority is:

6) Development of a high resolution, high efficiency gamma detector capable of stable ambient temperature monitoring.

There is also a need to develop infrastructure for rapid airborne and waterborne monitoring of noble gases and particulates and to develop a rugged system for gas sampling during On-Site-Inspections. While the panel notes that these are primarily systems development and implementation tasks, they facilitate many of the above radionuclide monitoring problems.

3.5 OTHER TECHNOLOGIES

The primary additional monitoring technology involves satellite systems that can monitor the optical, electromagnetic, and nuclear radiation from nuclear explosions, including x rays, gamma rays, and electromagnetic pulses (EMPs). Satellites also provide imaging capabilities that play a role in monitoring the underground testing environment. Much of the current system involves sensors that are deployed on the network of Global Positioning

[21] Source terms give the amounts of various diagnostic radionuclides likely to be released by explosions of different size, depth, and environment. See Appendix G.

System (GPS) satellites. DOE conducts a substantial research program in this area, and satellites play a key role in NTM monitoring of atmospheric and space environments. There will be no satellite data provided by the IMS, and the technical capabilities of the U.S. national system are sensitive so the panel does not address this area in detail. Although U.S. satellite capabilities are substantial, intrinsic limitations of overhead methods preclude sole reliance on this system for monitoring CTBT compliance in all environments. However, the capabilities provided by satellite systems must be assessed when evaluating the need for research on and improving the performance of other monitoring technologies. Potential synergies with other monitoring methodologies are discussed in the next section.

The research program supporting CTBT monitoring must also accommodate innovative new monitoring approaches beyond those of the well-established systems described in this chapter. For example, there is potential use of high-resolution GPS sensors to monitor ionospheric oscillations induced by large explosions. Also synthetic aperture radar may be used to monitor changes in ground surface in areas of concern that have been identified by other means (Meade and Sandwell, 1996). The basic mechanism by which the United States can sustain its technological edge and develop new creative monitoring strategies is the maintenance of a broadly based research program that is driven by verification needs, not constrained by existing operational perspectives; this is discussed further in Chapter 4.

3.6 OPPORTUNITIES FOR NEW MONITORING SYNERGIES

Each of the three technologies—seismic, hydroacoustic, and infrasonic—has primary and complementary roles. Seismic sensors are the best detectors of signals generated by underground events; seismic and hydroacoustic sensors are sensitive detectors of signals from events in water; and satellite and infrasonic sensors can monitor explosions in the atmosphere. However, energy couples from one medium to another and this coupling offers the opportunity for detection by multiple technologies. For example, explosions and earthquakes on land can generate hydroacoustic waves at continental margins, and upcoming seismic waves that strike the ocean floor can generate pressure waves in the water that are detectable by hydroacoustic sensors. Similarly, the surface motion from a shallow explosion on land or in the oceans can generate pressure pulses in the air that may be detected by infrasonic sensors. Conversely, explosions detonated at low altitudes in the atmosphere near the ocean's surface can generate signals that propagate underwater and could be detected by hydroacoustic sensors, and explosions in water can generate seismic waves that are detectable by seismic instruments on land (e.g., by T-phase stations).

Synergy in CTBT monitoring occurs at all stages of the monitoring process (i.e., detection, event association, location, identification) and often involves an interplay among them. For example, in the detection and association areas, joint association of hydroacoustic and seismic waves may define events that would not have been recognized by either of the technologies alone.

In other situations, the preliminary event definition may come from a single technology. The preliminary event location and origin time from this initial detection can focus the processing of data from other stations on limited time windows and azimuths of arrival. This focused, sensitive processing has at least two benefits. First, it allows the use of tuned processing techniques operating at lower signal-to-noise ratios. These techniques would generate an unacceptable number of spurious detections if used routinely. For example, hydroacoustic processing routinely operated with the lower signal-to-noise ratios used in focused processing would detect a large number of signals from explosions conducted for oil and gas exploration in the oceans. The resulting flood of detections would pose severe challenges to the association process, resulting in both incorrect associations and missed events. A related benefit of focused processing is that it can provide confidence in the monitoring of small signals that otherwise would have been undetected or detected but incorrectly associated with other signals. Conversely, focused processing as the result of a seismic detection may detect and identify small direct and reflected hydroacoustic phases associated with the preliminary definition of the event and thus strengthen confidence in the initial association. Another example of hydroacoustic and seismic synergy is found in the fact that characteristics of the hydroacoustic signals (i.e., the "bubble pulse") can provide definitive identification information for seismic events detected in the ocean.

The use of T-phase stations is yet another form of synergy between hydroacoustics and seismic waves. These seismic instruments, located on islands, record seismic waves generated by the impact of hydroacoustic waves. It is intended that the use of this phenomenon will provide additional coverage of the oceans and a greater possibility of detecting hydroacoustic waves from sources in remote or shadowed oceanic regions. The resulting detections will then be combined with detections from hydroacoustic stations in the association process. Note that the hydroacoustic signals from atmospheric events may be too small to be detected by the T-phase stations. Furthermore, the noise level at these stations is expected to be high. In general, the coupling between seismic and hydroacoustic waves is very poorly understood. For this reason, research on T-phase coupling is given the highest priority in the area of research synergy.

Still other forms of synergy involve infrasound, seismic, and hydroacoustic technologies in various forms. Processing data from the infrasonic arrays can be expected to provide constraints on the source locations based on azimuth determinations. However, the origin times are likely to be poorly constrained. The locations and approximate origin times can be used to focus the beams of seismic arrays and thus enhance their phase detection capabilities. If an event is located at sea, hydroacoustic processing can be focused on appropriate time windows. The interplay between seismic and infrasonic sensors provides yet another form of synergy in the detection process. Atmospheric pressure waves generated by near-surface chemical explosions (e.g., mining explosions) cause the surface of the Earth to move, and this motion is recorded by seismic instruments. These surface motions arrive after the seismic waves generated from the same source, appearing as lonesome waves that may be associated incorrectly with other phases or viewed as the only detected wave from a small event. Collocation of seismic and infrasonic sensors would indicate the coincidence of atmospheric waves and seismic motions, thus reducing the uncertainty of these detections. With this identification, seismic waves could be removed from further processing to reduce the number of phases involved in the association process. For stations near mining districts, this synergy of infrasound and seismology would be a significant benefit. Because estimates of the back-azimuths from infrasonic data can have smaller variances than those from seismic arrays, combined use of these data for event location would also be useful. In addition, near-surface mining explosions can be identified by associating seismic locations with infrasonic detections. However, such identification does not preclude masking, and the absence of radionuclide detections may be necessary to confirm that a surface explosion was not nuclear.

Events identified as suspicious on the basis of seismic, hydroacoustic, and infrasound observations may serve to prompt rapid response deployments of radionuclide equipment if such are developed for NTM. Forward predictions or radionuclide detections by permanent stations can also be made in this case using atmospheric transport models. Additional synergies between the technologies will emerge from experience with the complete international and national monitoring systems. Most of the synergies will depend on regional properties, and their definition and effectiveness will depend on calibration and operational experience in the region.

Many potential synergies exist between satellite methodologies and other approaches. For example, the development of ground truth data bases for each monitoring method could be augmented by coordinated use of overhead imagery. This has been established in the past for major nuclear test sites, where surface collapse craters could be associated with seismic event locations to calibrate the seismic array, eliminating biases due to unknown Earth structure. Future applications along similar lines could be pursued for calibration of quarry mines, earthquakes that rupture the surface, volcanic eruptions, and atmospheric disturbances (e.g., lightning). Overhead imagery has remarkable capabilities, but the need for focusing attention on a given region must come independently. For example, for On-Site Inspections, overhead imagery in a localized region can narrow down the region for assessment, but this must be guided by the initial location estimate from other methodologies. In the case where candidate testing sites are identified prior to an actual test, perhaps by seismically recorded quarry blasting of unusually large size in the area, an archive of overhead images could be compiled for comparison with images taken after a suspicious event is detected. This cannot be done with high resolution on a global scale, but it can be done for limited areas if there is a basis for defining those regions. Such approaches are still limited by the fact that underground testing does not have to result in detectable surface features such as collapse craters above the source. (The U.S. testing experience at Rainier Mesa has been that craters almost never were produced at the surface for these tunnel shots.) High-resolution, multispectral satellite imaging capabilities will become more available in coming years, which will make some aspects of satellite monitoring capabilities available to all nations. This availability should open up research opportunities on a broad scale.

Summary of Research Priorities Associated with Synergy

In summary, a prioritized list of research topics to increase the synergy between CTBT monitoring technologies includes:

1) Improved understanding of the coupling between hydroacoustic signals and ocean island-recorded T-phases, with particular application to event location in oceanic environments.
2) Integration of hydroacoustic, infrasound and seismic wave arrivals into association and location procedures.
3) Use of seismo-acoustic signals together with an absence of radionuclide signals for the identification of mining explosions.
4) Explore the synergy between infrasound, NTM, and radionuclide monitoring for detecting, locating, and identifying evasion attempts in broad ocean areas.
5) Determine the false alarm rate for each monitoring technology when operated alone and in conjunction with other technologies.

3.7 ON-SITE INSPECTION METHODS

The limited temporal duration of some of the effects associated with underground nuclear explosions (e.g., seismic aftershocks and radioactive gases) places a premium on rapid access to the site. The limited spatial extent of the anomalies and the limits on their detectability place a premium on the accuracy of the locations determined by remote sensors. Although aerial surveillance may serve to pinpoint some regions of interest and eliminate others, the 1000 km^2 location goal for remote sensors still leaves a large area to be covered by an On-Site Inspection. As stated before, improvements in the accuracy of locations determined by remote sensors are essential for effective OSI. A high priority for the overall OSI process is the elimination of significant systematic errors in the location capability of remote sensors, since these errors could completely negate the value of an On-Site Inspection. Reduction in the size of the random error is also important. However, the deterrence value of OSI may be preserved even if the random error is of moderate size as long as the systematic error is small.

Even if such improvements do occur, an effective OSI regime demands coupling an effective reconnaissance mode with rapidly deployable, efficient, focused operations at specific sites. These local operations employ relatively well-known geophysical technologies. What are needed are rapid deployment methods for sensitive instruments, criteria for evaluating the significance of locally collected data, and the signal processing ability to evaluate data quickly.

For OSI, radioactive gases leaking from cracks and fissures may be the only indicator of detonation location. Rapid noble gas monitoring equipment mounted in slow-moving aircraft may be required to perform the initial screening of an area to identify the approximate test location. Such air surveys may require the collection and rapid analysis of many air samples over a wide area. If the general area of radioactive noble gas release has been identified by airborne screening, surface vehicles can be used for sample collection. (e.g., Jester and Baratta, 1982; Jester et al., 1982).

In the broad, sense similar considerations apply for tests at sea. However, the limitations imposed by lack of IMS resources (e.g., ships, planes, and airborne gas and water samplers), the absence of long-lived deformation effects, and the possibility of decreased accuracy of location estimates for events in the ocean make it difficult to conduct an oceanic OSI. However, the possibility that currents may slowly disperse a relatively intense radioactive slurry from test debris may offer some hope of detection if seawater samples can be obtained from the proper locations. Finally, it should be noted that OSI at sea does not, in general, offer the same deterrence benefits as those on land. Its value could be enhanced if it were possible to attribute the nuclear device to a limited number of sources on the basis of chemical and physical properties of materials in the debris—another area of potentially fruitful research.

4

U.S. Research Infrastructure

INTRODUCTION

Pursuing the research described in the previous chapter will require a sustained federal effort, using the intellectual and technical resources of universities, private consulting companies, National Laboratories, and operational agencies. If the United States is to achieve the stated goal to confidently monitor evasively tested nuclear explosions with a yield of a few kilotons, additional fundamental and directed research will be needed. Simply deploying sensors and bringing the IMS and NTM into operation cannot guarantee technically effective monitoring of the treaty. Further research will enhance the reliability and performance of the CTBT monitoring effort and should lead to cost reductions as the dependence on human analysts decreases and erroneous identifications and On-Site Inspection requests are avoided. The technical knowledge resulting from this research program, from its integration with the efforts of the U.S. NDC and its connections with other agencies and organizations carrying out related research, will also be valuable for addressing claims of CTBT violations made by other nations based on IMS data. In addition, it will benefit other programs of U.S. national interest.

To illustrate the programmatic challenge of the research effort, this chapter describes past and present research programs in support of test ban monitoring. First the chapter establishes the current baseline for support of basic research programs related to CTBT monitoring. The panel then concludes that the current levels are small compared to the needs for supporting the priority research topics in Chapter 3. Consequently, increased funding for basic research will be required to support U.S. efforts to meet national monitoring goals. This chapter also discusses the panel's view of the characteristics of an effective long-term research program and the mechanisms to transition research results to an operational environment. Additional details about the past seismic research programs are provided in Appendix B.

4.1 STRUCTURE OF CURRENT DoD AND DOE PROGRAMS

Sections 2.6 and 2.7 have provided an overview of the research structure that is presently in place to support CTBT monitoring. Of particular importance for assessment of the future program structure is the reorganization in DoD programs that took place in 1996 (see Section 2.7). The consolidation of separate lines of funding for CTBT-related research that had existed in the Air Force

Office of Scientific Research (AFOSR), Air Force Phillips Laboratory (AFPL), and ARPA research budgets eliminated the DoD distinctions between basic, applied, and development research programs (i.e., the former 6.1/6.2/6.3 designations). Thus, the programmatic structure and balance to the research effort has been removed and a new approach must be established. The new system should service the immediate operational needs of the NDC and provide the mid- and long-term improvements necessary to maintain confidence in the verifiability of the CTBT. It must do so in the face of changing world situations and a background of numerous nonnuclear events with characteristics similar to nuclear explosions. Meeting these challenges requires a properly organized and managed research program that includes fundamental research conducted by universities, applied and developmental research conducted by private companies, and advanced systems development efforts provided by private companies. These efforts must be coordinated with the efforts and needs of other agencies involved in CTBT monitoring and still other agencies and organizations that use similar technologies.

The FY97 Department of Defense (DoD) external research program is presently funded at a level of $8.8M. In FY 1997 DOE will provide about $0.4M for external research.[22] The DoD FY97 budget for development of the IDC system is about $20M, which will need to be sustained at least until the IMS system is deployed. The funding level for basic and applied research for CTBT monitoring from all sources was around $12 million in FY 1995 and FY 1996.[23] If DOE support of external CTBT-related fundamental research decreases as indicated in its current budget projections, the remaining $8.8 million for DoD programs represents about a 27 per cent reduction of the research effort, despite the expanded role of multiple monitoring technologies and the time frame established by the signing of the CTBT in 1996. (At this time, it is difficult to project DOE's effort into the future, largely due to turnovers in management personnel; however, no explicit commitment to sustaining an external research program anywhere near the scale of the FY 1995 and FY 1996 programs has been made.)

Past programs have supported a broad range of research that should be sustained to enhance U.S. CTBT monitoring capabilities. Examples of basic and applied research projects that were supported by AFOSR, AFPL, AFTAC, ARPA, and DOE in FY 1995 and 1996 are summarized in Lewkowicz et al. (1996). These papers describe studies of detection, location, and identification of seismic sources; hydroacoustic and infrasonic wave propagation; radionuclide monitoring; and data processing and analysis. Given the progress to date, it is clear that the challenge of effective CTBT verification requires expansion of this type of research to support effective monitoring of all testing environments (other than space) in the near, mid-, and long terms.

If the United States is to meet its stated monitoring goals in a timely manner, substantially higher funding than the combined DoD-DOE budget of $12 million/yr for the previous two years will be needed to support the high-priority research issues listed in Chapter 3. The panel's conclusion on the need for increased funding levels is based on the following factors: 1) the geographic breadth of the areas of monitoring interest, 2) the expanded range of monitoring technologies that will be incorporated into the IMS, 3) the lack of previous monitoring experience and global data for some of the new monitoring technologies, 4) the need to carry out new fundamental research to explore the synergies between monitoring technologies, and 5) the importance of basic research programs for training technically competent scientists who will participate in U.S. monitoring operations.

The basic research program being considered is separate from the systems development effort directed at setting up the IDC and also from internal DOE and DoD budgets that support both applied and advanced developmental research. Some proposed undertakings, such as a long-term committed seismological effort to developing a detailed model of the crust and upper-mantle velocity structure under the major land masses of

[22] By comparison, DOE had provided $4 M/yr from the FY95 and FY96 budgets to external research. These compare with the total FY97 DOE CTBT budget of $69.6M.

[23] This does not including systems development efforts for the NDC, efforts related to satellite monitoring, systems development efforts directed at setting up the IDC, and internal DOE and DoD budgets that support both applied and advanced developmental research.

interest early in the twenty-first century, would require higher funding levels but could result in optimal operational capabilities at that time. Similarly, development of portable rapid radionuclide processing systems could be undertaken if it is decided that this capability is needed, which would also require additional funding levels. Some large-scale field experiments involving controlled explosions and field recordings would likewise require more substantial budgets.

The eventual funding level for R&D programs in CTBT monitoring presumably must be determined after assessments by the Administration and by Congress of the importance of reaching U.S. monitoring goals. The linkage between R&D and operational improvements in monitoring is explicitly recognized in President Clinton's statement of August 11, 1995. Given the CTBT safeguards, it is clear that, at a minimum, current levels of support should be sustained. However, the discussion in Chapter 3 suggests that incremental improvements will not achieve U.S. monitoring objectives, and research is the most likely activity to achieve breakthroughs in monitoring capabilities. If the capabilities of the IMS do not improve with time, there may be false alarms associated with erroneous interpretations of the monitoring data. Such events would be costly if they resulted in an OSI.

Deployment of IMS seismic, hydroacoustic, infrasonic, and radionuclide stations will expand opportunities for data analysis and research progress in a variety of applications other than CTBT monitoring that are now constrained by data availability. If IMS data become available, it will be a valuable contribution to CTBT monitoring if agencies such as NSF, USGS, and NOAA fund research related to multiuse applications of the data stream.

4.2 RESEARCH PROGRAM BALANCE

Several important issues relate to program balance for U.S. CTBT research. These include the relative levels of disciplinary efforts; the relative emphasis on fundamental, applied, and advanced developmental research; and mechanisms to coordinate and buffer the competition for resources that can develop between different types of research. The 1995-1997 CTBT research program is the most useful basis for projecting future needs, given that prior to 1995, relatively little emphasis in external fundamental research programs was placed on monitoring technologies other than seismology. For example, as noted above, the total DOE CTBT research program (most of which is for research at four National Laboratories) involved much more support in seismological research ($11.6 million), compared to infrasound research ($1.1 million), hydroacoustic research ($1.2 million), and radionuclide instrumentation development ($3.4 million). The $12 million per year external research programs administered by AFOSR and AFPL for FY 1995 and FY 1996, with support derived from DOE, AFOSR, ARPA, AFTAC, and ACDA, had a similar emphasis on seismology, with modest proportionally smaller efforts in fundamental infrasound, hydroacoustics, and radionuclide research.

Four factors influence the higher priority that has been assigned to seismological research in past programs: (1) seismology has been well suited for monitoring the underground nuclear testing programs of the former Soviet Union, China, and France; (2) seismic monitoring on a global basis presents a well-defined challenge that must be met if U.S. CTBT monitoring objectives are to be achieved; (3) there have been no data sets or CTBT-focused research efforts in the other disciplines, and IMS data are just beginning to accumulate and to clarify operational challenges; and (4) some of the additional technologies are complementary to other NTM. For example, for the United States, infrasound is deemed largely to be a backup system for satellite monitoring as far as atmospheric explosions are concerned. As a result, throughout the U.S. technical discussions during CTBT negotiations, hydroacoustic, radionuclide, and infrasonic monitoring challenges have been considered less profound than those for seismic monitoring. It is important to recognize that the dependence on complementary monitoring capabilities and synergies has resulted in relatively minimal capabilities for some systems (e.g., the IMS hydroacoustic network), making it essential to exploit those synergies optimally within the context of the overall capabilities.

The recommendation above for an increased funding level pertains to the external basic and applied research effort, which is only one element of the overall CTBT research effort. The FY 1997 DoD budget for development of the IDC is about

$20 million, and is clearly a high priority to sustain this effort until the IMS is operational. Some projects funded from this budget line are considered research, but it is a different "flavor" of research from the basic and applied efforts emphasized in Chapter 3. The development and implementation of communications and computer hardware and the development of automated processing software are essential for the IMS and U.S. NDC but will not solve the basic questions that arise in analyzing data from various monitoring systems. Future CTBT research must sustain this type of advanced developmental effort, but it is important to ensure that time urgency and high costs do not squeeze out the fundamental science research program on which future improvements depend so strongly. There was a tendency for this to occur in the former ARPA program, and the new DoD program will have to establish effective firewalls to prevent it. When the fundamental research program budget experiences large fluctuations there is great difficulty in sustaining steady progress on the many research fronts that exist, and the instability can cause university researchers and private companies not to commit to relevant research efforts. This instability means that the best researchers will turn away from CTBT monitoring and take up other fields.

Other research efforts now conducted within government agencies are also separate from the external fundamental research program and should be sustained. For example, the large applied and advanced developmental program of DOE, which addresses all areas of relevance to CTBT monitoring including satellite NTM and On-Site Inspection procedures, should be sustained with budget levels similar to FY 1997 since essential IMS-NTM regionalization and calibration activities are being pursued. Research on atmospheric transport models has been supported for years at NOAA and by DOE at the National Laboratories. The importance of these models for radionuclide backtracking and infrasonic monitoring motivates sustained research in this area. If possible, university research in atmospheric transport modeling should also be encouraged. Instrumentation deployment and operations efforts performed and funded by AFTAC, NSF, and the USGS are also providing essential parts of the IMS and NTM and should be funded with appropriate budget levels. It is important to recognize that government laboratory research programs tend to emphasize applied research, which—if successful—meets some identified operational needs. University and private company projects tend to be less tightly constrained by operational needs and consequently are able to explore areas of research that may have no immediate application but could lead to totally new approaches. The distinctions are not hard and fast, and both groups do research of all types.

With diverse research efforts spread across universities, private companies, and government agencies, it is important to have good coordination of CTBT research. Consolidation of DoD research programs provides an immediate opportunity to enhance such coordination within DoD (at one time it was effectively provided by ARPA). In addition, it offers an opportunity to coordinate with other government and nongovernment agencies doing research and development relevant to CTBT monitoring. DoD and DOE have memoranda of understanding, which effectively bridge the DOE research effort to AFTAC operations, but there appears to be room for improvement in this coordination as well. A better flow of information and assessment of capabilities, upward from the operational environment to the policy and intelligence arenas (ACDA, State Department, and intelligence organizations) and outward to investigators familiar with the various technologies are also important. At present there is little coordination between the explosion monitoring programs of DoD and DOE and the earthquake monitoring programs of the USGS. This is unfortunate because the signals requiring the greatest analysis efforts at the IDC will be earthquake seismograms, often derived from source areas where relevant information can be obtained via the USGS. The USGS also has a key role in the documentation of U.S. seismicity—providing additional data to interpret signals from U.S. blasting and earthquakes, that may be recorded by the IMS and be of some foreign concern. This is an example of an important benefit that would be provided by multiuse of IMS data for research purposes.

4.3 COORDINATION WITH OTHER NATIONAL AND INTERNATIONAL EFFORTS

Most of the technical disciplines that contribute to CTBT monitoring also contribute to independent activities of national concern and, therefore, have support from other government agencies. For example, seismology is the cornerstone of earthquake hazard assessment, as well as having a primary role in basic science investigations of the Earth system. Hydroacoustics has long been a mainstay of submarine monitoring and recently has emerged as a potential technology for monitoring global warming. Technical advances in these areas may often benefit CTBT monitoring (e.g., new three-dimensional velocity models or improved wave propagation modeling capabilities in the oceans and solid Earth), and the CTBT research program needs to exploit and possibly coordinate with such efforts. If the IMS data is provided to the scientific community, dual-use applications using these datastreams (e.g., for studies of natural hazards) will help to sustain the long-term viability of the CTBT monitoring system. Some of the multiuse applications of seismic data were discussed in NRC (1995). There are also many opportunities for multiuse of the hydroacoustic, infrasonic, and radionuclide data.

Specific research efforts that the CTBT program could pursue in coordination with other agencies include development of improved three-dimensional Earth models for areas of CTBT monitoring interest or a focused effort to determine the gross structure of the crust and lithosphere in Eurasia and other continental areas of interest. These are topics of relevance to global earthquake monitoring and basic science investigations of the Earth supported by NSF and the USGS. There are also fundamental topics such as improved earthquake location methods that span many applications, and the coordination of national research efforts could accelerate progress in these areas and eliminate unnecessary redundancy.

Voluntary international data exchange on calibration events is one of several confidence-building activities discussed in the CTBT. Precisely located earthquakes and details about large mining explosions can greatly accelerate seismic calibration of a region, and it is desirable to pursue such exchanges of data and the compilation of associated ground truth data sets for calibration of the monitoring technologies.

4.4 REQUIREMENTS FOR LONG-TERM STABILITY AND EFFECTIVENESS

The most important requirement for stability of the CTBT research program is stabilization of the research budget, with a multiyear commitment that firmly establishes its viability for intellectual resources in universities and private companies. Such stability is essential for training technically competent scientists and researchers who will participate in U.S. monitoring operations. Without it, bright young researchers will not enter the fields supportive of CTBT monitoring. To the extent possible, the research program should be buffered from fluctuations imposed by systems development and operational emphases. It should, however, include effective communication of operational needs to the research community.

It is well established that the quality of research programs is enhanced by using peer-review systems, and this is desirable for the CTBT fundamental research program as well. Although some applied and most advanced developmental activities can best be pursued with focused Requests for Proposals (with responses being assessed by government program managers), more flexible announcements for basic research funding need to be made in the external fundamental research program to ensure the influx of innovative ideas and creative approaches to established research areas. Peer review ensures a healthy program that captures cutting-edge approaches and avoids entrenchment. The lack of DSWA experience in supporting basic research programs in universities is a concern, and the lack of a clear distinction between basic and applied research complicates the maintenance of a long-term program.

Annual multidisciplinary workshops involving all of the IMS technologies would provide an important mechanism for communicating research advances and operational needs. Such meetings would also contribute to the development of synergistic monitoring strategies by promoting communication between different research communities. Publication of a comprehensive workshop report would also be a valuable part of this effort. The panel notes that similar workshops were supported by the previous AFOSR program. The last two meetings involved all of the IMS monitoring technologies.

In recent years, AFTAC has enhanced its internal technical expertise by hiring several well-trained seismologists, even when confronted with mandated reductions in manpower. This has proved effective in accelerating the incorporation of new research advances and complex analysis procedures into the operational regime of the U.S. NDC. Increasing the number of Ph.D.-level scientists at the NDC is a proven strategy for abetting technology transfer and is supported by the panel.

An additional means by which research efforts can be transitioned effectively to the operational environment would be the establishment of a CTBT research test bed. Assuming that there is open access to the IMS data, this would require a facility that replicates significant aspects of the IMS and U.S. NDC monitoring system. The facility would have real-time data processing capabilities and historical data archive access. The prototype IDC has operated a limited system of this type, with visitors to the Center for Monitoring Research accessing the IMS data and processing system, but at present there is no clear plan for a broadly accessible test bed system for the long term. Progress on many of the research issues raised in Chapter 3 will require researchers to analyze actual signals from various monitoring disciplines and give them an opportunity to test proposed analysis methods. This analysis and testing would cultivate the development of new methods in a software environment that is much closer to the operational situation than currently exists at any university or private company, as well as providing realistic constraints on processing. A test bed facility, preferably operated by the U.S. NDC, could also serve as a site for focused investigations of problem events in which experts gather to address technical issues (either in person or by computer linkup). This could be designed to be responsive to both short-term and long-term problems that arise in the operational arena. Finally, such a test bed could form the basis for regularly scheduled exchanges between the policy and technical communities, which would make clear the processes, constraints, and uncertainties under which both communities operate. One possibility would be to have the prototype IDC transition into this type of test bed facility once the permanent IDC is established in Vienna. This transition would require substantial funding beyond that described in the basic research budget above.

5

Conclusions And Recommendations

5.1 CTBT MONITORING CHALLENGES

The physical processes associated with nuclear explosions produce distinctive sources of acoustic waves, elastic waves, radiation, and radioactive materials and gases. These signals and products then propagate through or are advected by the Earth system with various transmission effects, and may eventually be detected by different types of sensors placed around the planet's surface or on satellites. The types of signals that can be recorded and interpreted are limited by the extensive background noise of the Earth (e.g., earthquakes, weather phenomena, conventional explosions), and the physical limitations of the sensors (e.g., bandwidth, sensitivity). Signals recorded at different locations must be retrieved from the field, associated with a time and location for a common source using general knowledge of signal propagation. Attributes of the recorded signals, corrected for propagation and instrumentation effects, are then used to identify the type of source, ideally distinguishing nuclear explosion signals from earthquakes or other non-nuclear phenomena. All monitoring technologies share these fundamental elements; source excitation, signal propagation or advection, recording instrumentation, event association, event location, and event identification. They also share the technological challenges of data retrieval and automation of data analysis.

Any CTBT monitoring system will have practical limits in the capabilities to detect, locate, and identify events based on the analysis of the recorded signals. These limits are imposed both by cost considerations that constrain the data acquisition and processing and by intrinsic constraints of the monitoring technologies. A complete interpretation of the monitoring limits must allow for the possibility of various evasion approaches, such as muffling the nuclear explosion signals by detonation of the device in a pre-existing cavity (decoupling) or obscuring the explosion signal by simultaneous detonation with an earthquake, quarry blast, or mine collapse. More than 50 years of research underlies the present ability to use the various wave types in diverse environments for monitoring applications. The significant progress that has been achieved has provided the technical basis for moving forward with the CTBT negotiations. However, the national objectives for assuring international compliance with a total ban on nuclear explosions place extreme demands on all of the monitoring technologies and operational systems, and there is a need for continuing research to enhance the entire U.S. CTBT monitoring system.

It is clear from the CTBT negotiating record that a monitoring capability of "a few kilotons, evasively tested" in selected areas of the world is the goal for U.S. monitoring capabilities.[24] The primary technical challenge associated with the CTBT is related to the fact that even very small tests are banned. In this setting, decreasing the magnitude threshold for the monitoring goals has several important implications for the performance of a monitoring system: the overall number of detected events increases sharply, the number of stations in a fixed network capable of detecting a given event decreases, and the distance at which detections can be made decreases. Given these factors, the detection, location and identification of small events by the combined IMS and NTM assets involves the analysis of signals recorded at regional distances where signal propagation is often complicated and regionally varying. Even with only a few reporting stations, a monitoring system needs to provide high confidence locations with an uncertainty smaller than 1000 km^2 for on-land events. This requirement reflects operational requirements mandated by the On-Site Inspection provisions of the treaty (Protocol to the CTBT, Part IIA). Given these requirements, the challenge of precisely locating and confidently identifying all small events at some low magnitude threshold given sparse monitoring networks is formidable. Meeting it requires a sustained basic research program in support of CTBT monitoring.

Political realities mandate a long-term commitment by the U.S. government to monitor international compliance with the CTBT, using cost-effective, highly reliable technologies. Present technologies cannot achieve the highest levels of confidence at very low yields, and this prompted President Clinton to call for "pursuing of a comprehensive research and development program to improve our treaty monitoring capabilities and operations" as one of six CTBT Safeguards. Thus, a sustained basic research program is required to enhance the performance of the U.S. CTBT monitoring system. This report identifies key areas of research that will contribute to achieving the national monitoring goals: the disciplines of seismology, hydroacoustics, infrasonics, and radionuclides, all elements of the U.S. National Data Center and the International Monitoring System

The research program that will best serve the national needs will sustain both long-term and short-term efforts, and will span the spectrum from innovative exploratory research to advanced developmental efforts. It will draw upon expertise in universities, private industry, and U.S. National Laboratories. The external research program supporting non-governmental fundamental research should involve a funding level in excess of the current commitment ($8.8M). This is in addition to the internal programs of DOE and DoD and the developmental research that supports the IDC and NTM. There should be close coordination between DoD and DOE elements of this program, and strong integration with the operational effort conducted by the U.S. National Data Center operated by the Air Force Technical Applications Center, on-going interactions with the policy community, and cooperation with agencies using similar technologies to address challenges of national interest. The panel emphasizes that open access to IMS data would facilitate this cooperation that would be of great benefit to CTBT monitoring.

5.2 RECOMMENDATIONS

The panel elaborates on these issues in response to the elements of its charge.

What are the basic research problems remaining in the fields of seismology, hydroacoustics, infrasonics and radionuclides that should be pursued to meet national and international requirements for nuclear monitoring? The panel's work on this question should anticipate quality of data to be made available in the future, in particular those data from the CTBT International Monitoring System.

The United States has 5 primary technical CTBT monitoring methodologies available to it: seismology, hydroacoustics, infrasound, radionuclide, and satellite systems. All have mature theoretical development, advanced recording instru-

[24] The Geneva working paper CD/NTB/WP.53 of 18 May 1994 stated the U.S. position that: "The international monitoring system should be able to .. facilitate detection and identification of nuclear explosions down to a few kilotons yield or less, even when evasively conducted, and attribution of those explosions on a timely basis."

mentation, and efficient data collection, but they differ significantly in the specific technical challenges that arise for CTBT monitoring and the amount of prior global monitoring experience that is available for them. The first four of these form the basis of the IMS capabilities. Recommended research areas for these disciplines are described below.

Seismology

Seismological monitoring is an advanced and mature discipline in many areas related to CTBT monitoring. For decades, nuclear testing treaties have been verified using seismic monitoring of teleseismic signals (i.e., signals that are recorded at distances larger than 2000 km from the source). These signals have relatively simple propagation effects that are now well understood. Using such data, global detection, location, and identification of all underground events above magnitude 4.5 appears to be straightforward given an adequate distribution of recording facilities. As now planned, IMS and NTM assets will meet this requirement.

Teleseismic signals are weak, however, for the small events of interest for CTBT monitoring (including events with magnitudes as low as 2.0 in some regions). Consequently, treaty verification will necessitate increased dependence on regional signals of small events, observed at distances less than about 1000 km. These signals are complicated by reverberations in the crust, but they often have good signal-to-noise ratios.

Pushing the seismic monitoring magnitude threshold downward to include precise event location and high confidence identification for small events is the primary motivation for continued seismological research. For this reason, there is need for research on all aspects of detecting, locating and identifying events in the magnitude range 2.0-4.5 using regional signals from known sources in diverse regions of the world. For this task, the IMS and NTM seismic stations need to be carefully calibrated for location and magnitude determinations using regional distance observations in various parts of the world. The many other seismic stations that exist outside of the NTM and IMS systems can be used to determine crustal properties, ground-truth event parameters, and development of innovative analysis procedures. In addition, research on the characteristics of seismic radiation from small events and seismic wave propagation in the heterogeneous crust of the Earth should be conducted. This will improve the capability to identify small nuclear explosions amidst a background of numerous small earthquakes and quarry blasts. A prioritized list of research activities in support of seismic monitoring includes:

1) Improved characterization and modeling of regional seismic wave propagation in diverse regions of the world.
2) Improved capabilities to detect, locate, and identify small events using sparsely distributed seismic arrays.
3) Theoretical and observational investigations of the full range of seismic sources.
4) Development of high-resolution velocity models for regions of monitoring concern.

Hydroacoustics

Monitoring sound waves in the oceans is a well-advanced discipline, primarily as a result of investments in Anti-Submarine warfare. The ocean medium is a remarkably efficient transmitter of low frequency acoustic waves, so that even modest conventional explosions in most regions of the deep oceans are readily detected and identified with adequate instrumentation. Because hydroacoustic waves also couple efficiently into seismic waves at the ocean bottom (and vice versa), the medium can be effectively monitored by a combination of hydroacoustic and seismic networks.

To date, there has been relatively little research on the use of hydroacoustic signals to monitor underground and atmospheric explosions. Given that the proposed IMS hydroacoustic network will use a small number of sensors, with no directional capabilities, the panel concludes that the system will have extremely limited detection and location capabilities. Using only hydroacoustic data, it will be difficult to reduce false alarms from natural and human sources and to identify and locate sources in shallow or polar regions. Because of these deficiencies, there is a need for research on synthesizing hydroacoustic data with seismic, infrasonic, and radionuclide information and to assess the capability of the integrated system to monitor within national goals. Prioritized research topics in support of hydroacoustic monitoring include:

1) Improvements in source excitation theory for diverse ocean environments, particularly for earthquakes and for acoustic sources in shallow coastal waters and low altitude environments.
2) Understanding the regional variability of hydroacoustic wave propagation in oceans and coastal waters and the capability of the IMS hydroacoustic system to detect these signals.
3) Improved characterization of the acoustic background in diverse ocean environments.
4) Improving the ability to use the sparse IMS network for event detection, location, and identification and developing algorithms for automated operation.

Infrasound

While several atmospheric nuclear tests were conducted after negotiation of the Limited Test Ban Treaty in 1963 (by non-signing nuclear weapon states, China and France), the reduction of such tests relative to the prior two decades brought about a decrease in atmospheric sound wave monitoring efforts by the United States This trend will be reversed by the IMS system which will establish a global array of infrasound sensors to enable routine monitoring of low frequency sound waves on a global basis for the first time in decades. At present, however, the United States has only a few experts in infrasound, and virtually no infrastructure for research in atmospheric monitoring using sound waves. Thus, the primary research issues associated with CTBT monitoring involve first-order questions about the background noise, involving wind noise reduction and the nature and frequency of events such as volcanic explosions, meteor impacts, sounds radiated from ocean waves (microbaroms), auroral infrasonic waves, and mountain associated waves. Research on these questions would be augmented by publication of basic information on the U.S. monitoring experience in the 1950's and 1960's. This would enable a wider understanding of likely infrasound signal strength for explosions of different yield, different environments, and different distances from sensors. To support and enhance the monitoring capabilities of the IMS infrasound network, the following are priority research topics:

1) Characterizing the global infrasound background using the new IMS network data.
2) Enhancing the capability to locate events using infrasound data.
3) Improving the design of sensors and arrays to reduce noise.
4) Analyzing signals from historical monitoring efforts.

Radionuclides

Radionuclides released from a nuclear explosion are distinct from nuclear reactor emissions and natural background radioactivity. Because radionuclide analysis can provide unambiguous evidence of a nuclear explosion, the IMS will receive data from a global network of fixed particulate and noble gas detectors. The data from this network will differ from the other monitoring technologies in two important respects. First, the raw data streams will consist of daily gamma-ray spectra for samples of wind transported gases and particulates, rather than the time series of seismic, hydroacoustic, and infrasound data. Consequently, there will be significant challenges in merging the analysis of radionuclide datasets with the other components of the IMS system. Second, and most important, the radionuclide network requires time scales as long as 10 days to two weeks to detect a possible nuclear explosion. This delay is sensitive to the rates of wind-borne transport of radionuclide particulates and the integration times for radiochemical analyses. Once an event is detected, further analyses of wind and climate patterns are required to back-track the data to locate the site of an explosion.

Given these limitations, a wide range of research is needed to strengthen the capabilities of radionuclide monitoring. Improvements are needed in the understanding of source terms.[25] and the airborne transmission effects. The source issues involve characterization of radioactive emissions from past nuclear tests (atmospheric, underwater,

[25] Source terms refer to the amounts of diagnostic radionuclides likely to be released by explosions of different sizes in diverse environments. See Appendix G.

and underground), with an emphasis on understanding atmospheric rain-out and underground absorption. Research is needed to study the various atmospheric effects associated with the dispersal of radioactive materials to improve the atmospheric transport models used for back-tracking of airborne radionuclides. In addition, there is a need to develop new instrumentation, infrastructure, and procedures for rapid radiochemical measurements. The goal of these efforts is to reduce the time delay between potential explosions and radionuclide detection, and to facilitate the work of On-Site Inspection teams. Priority research topics include:

1) Research to improve models for back-tracking and forecasting the air borne transport of radionuclide particulates and gases
2) Research and data survey to improve the understanding of source term data.[26]
3) Understanding of atmospheric rain-out and underground absorption of radionuclides from nuclear explosions.
4) Assessment of the detection capabilities of the IMS radionuclide network.
5) Research on rapid radiochemical analysis of filter papers.
6) Development of a high resolution, high efficiency gamma detector capable of stable ambient temperature monitoring.

What research is necessary to strengthen the synergy between the seismic, hydroacoustic, infrasonic, and radionuclide data sets to improve overall monitoring capability and to meet national and international requirements?

There are great opportunities to enhance the synergies between the different CTBT monitoring technologies. Energy propagating in the Earth system can couple from one medium to another (air to water, air to land, or land to water). Each monitoring technology has primary capabilities for sources in a particular medium as well as complementary roles for sources in the other media. For example, explosions and earthquakes on land can generate hydroacoustic signals when their seismic waves strike the continental boundaries or come up under the ocean bottom and convert to sound waves in the water. Possibilities for synergies in the use of diverse wavetypes exist in all stages of the monitoring process. Priority multi-disciplinary research that will enhance the synergy of monitoring technologies include:

1) Improved understanding of the coupling between hydroacoustic signals and ocean island-recorded T-phases, with particular application to event location in oceanic environments.
2) Integration of hydroacoustic, infrasound and seismic wave arrivals into association and location procedures.
3) Use of seismo-acoustic signals together with an absence of radionuclide signals for the identification of mining explosions.
4) Explore the synergy between infrasound, NTM, and radionuclide monitoring for detecting, locating, and identifying evasion attempts in broad ocean areas.
5) Determine the false alarm rate for each monitoring technology when operated alone and in conjunction with other technologies.

In addition, research synergy would be promoted by the communication of research advances and operational needs at annual multidisciplinary workshops.

How should the research results be transitioned so that they are most useful to those responsible for monitoring and verifying a CTBT?

Continuing basic research must be accompanied by effective mechanisms for transitioning research advances from academia and industry into the operational environment and for making the operational needs known to the research community. Given the continuing need for innovative exploratory research, advanced developmental research and operational advances, and realistic funding projections, the panel recommends that external research programs of DoD and the DOE should be carefully coordinated so that a balance of basic and applied efforts can be sustained. It is important to buffer fundamental research efforts from short-term operational needs, otherwise creativity and innovation will be curbed. At the same time, it is important that the research community be aware of the potential applications of

[26] Source terms give the amounts of various diagnostic radionuclides likely to be released by explosions of different size, depth, and environment. See Appendix G.

their work. The CTBT research program should adopt a hierarchical research infrastructure consisting of a broadly-based basic research program, overlain by increasingly focused applied research efforts that develop and support the transition of promising technologies into the operational environment. Past administrative subdivisions of basic, applied and advanced developmental efforts have not worked efficiently in the Air Force test ban treaty monitoring program, in part due to fluctuations and uncertainties in the budgets of the separate efforts. The CTBT research program requires more effective oversight, coordination, and funding stability than have existed for the last decade.

The panel concludes that increased numbers of Ph.D. level research staff at the U.S. National Data Center would help to promote technology transfer to the operational regime. Technical training and sophistication is essential for recognizing and rapidly incorporating research advances into operational systems.

An additional means by which research efforts can be effectively transitioned to the operational environment would be the establishment of a CTBT research test bed. This would require a facility that replicates significant aspects of the IMS and U.S. NDC monitoring system, with real-time data processing capabilities and historical data archive access. The prototype-IDC has operated a limited system of this type, with visitors to the Center for Monitoring Research accessing the IMS data and processing system, but there is at present no clear plan for a broadly accessible test bed system for the long term. Progress on many of the research issues raised in Chapter 3 will require researchers to analyze actual signals from the various monitoring disciplines, and have an opportunity to test proposed analysis methods. This would cultivate the development of new methods in a software environment that is much closer to the operational situation than currently exists at any university or private company, as well as providing realistic constraints on the processing. A test bed facility could also serve as a site for focused investigations of problem events in which technical experts gather to address technical issues (either in person or by computer link-ups). This could be designed to be responsive both to short-term and long-term problems that arise in the operational arena. Finally, such a test bed could form the basis for regularly scheduled exchanges between the policy and technical communities that would make clear the processes, constraints and uncertainties under which both communities operate. One possibility would be to have the prototype-IDC transition into this type of test bed facility once the permanent IDC is established in Vienna. This would require substantial funding beyond that described in the basic research budget above.

What are characteristics of a long-term program that would provide a stable, but adaptable base of support to those responsible for monitoring and verifying a CTBT?

The FY97 Department of Defense (DoD) external research program is presently funded at a level of $8.8M and DOE will provide about $0.4M ($4M/yr had been provided from FY95 and FY96 budgets) to external research out of the total FY97 DOE CTBT budget of $69.6M. The DoD FY97 budget for development of the IDC system is about $20M, which will need to be sustained at least until the IMS system is deployed.

If the United States is to meet stated national monitoring goals in a timely manner, increased funding (compared to the above budget levels) will be needed to support the high-priority research issues listed in Chapter 3. Pursuing the panel's recommendations on priority research will also require close coordination between several agencies. This effort will include focused missions to:

1) develop a new generation of Earth event catalogs in selected regions, listing all natural phenomena and human related sources down to magnitudes that are significantly below present-day capabilities;
2) utilize the data from the new global network of infrasound and radionuclide monitors;
3) improve global seismic velocity models;
4) integrate seismic and infrasound data to support the limited hydroacoustic network;
5) quantify the effect of seasonal variability on the location and detection capabilities of the hydroacoustic network; and
6) coordinate efforts to understand the structure of the crust and lithosphere in areas of interest to the United States at a level that allows reliable event identification.

The decision about whether to fund and pursue these research efforts will be influenced by assessments of the adequacy of CTBT verification and the benefits of these research undertakings.

To strengthen the CTBT research program it is also important to stabilize of the research budgets, with a multi-year commitment that firmly establishes the viability of this research area for intellectual resources in universities and private companies. This stability is essential for training technically competent scientists and researchers who will participate in U.S. monitoring operations and provide assessments of technical issues, problem events, and erroneous claims made by other nations. Without it, bright young researchers will not enter the fields supportive of CTBT monitoring. This research program should be buffered from fluctuations imposed by systems development and operational emphases, but should be run with effective communications of operational needs to the research community.

Research in the field of seismology is largely driven by the large numbers of non-nuclear events (earthquakes and conventional explosions) whose signals must be discriminated from those of potential nuclear explosions. Research challenges in the fields of hydroacoustics, infrasound and radionuclide monitoring will become more focused with the operation of the CTBT monitoring system. The panel notes that there is limited research support for some of the monitoring technologies outside of the CTBT research program, particularly for infrasonics and hydroacoustics. The rationale for increasing the current research program is that there are major unsolved problems in seismology, and that there will soon be a substantial flow of data from infrasound, radionuclide and hydroacoustic systems for which there is far less operational experience. Calibration of these systems involves both experimental and research issues, and support for university and contractor programs is vital to establishing a pool of national expertise in the analysis of these data, both for national CTBT monitoring activities and for competent assessment of claims that may be made by other countries based on observations from these technologies. At the same time, the balance of effort needs to reflect the role that these systems play in the overall U.S. capability, including satellite assets.

Additional conclusions regarding the stability and effectiveness of a CTBT research program include the following:

1) The 1996 consolidation of the DoD research program into a single program structure (now administered by the Defense Special Weapons Agency (DSWA) for the Nuclear Treaty Program Office (NTPO)), provides an unprecedented opportunity to form DoD's program into an effective CTBT research effort. This program should be structured to accommodate both long-term and short-term activities, as well as wide-ranging basic research and focused developmental research, if it is to prove effective in supporting the CTBT monitoring effort.
2) The panel concludes that the quality of research programs will be enhanced by the use of peer-review systems. Some applied, and most advanced developmental activities, can best be pursued with focused Requests for Proposals (with responses being assessed by government program managers). However, more flexible announcements are required for the external basic research program to ensure influx of innovative ideas and creative approaches to established research areas. Peer-review, involving scientists who are both scientifically and programmatically knowledgeable, ensures a healthy program that captures cutting edge approaches and avoids entrenchment.
3) An effective national program requires close coordination of the DoD program with DOE and the operational effort at the National Data Center, which is run by the Air Force Technical Applications Center (AFTAC). It is also important to sustain strong lines of communication with research programs in other agencies (such as the USGS, NSF, and NOAA) which provide basic and applied research advances and even operational products (e.g. precise earthquake bulletins and weather pattern models) that can augment CTBT monitoring. Such coordination would be enhanced if there is open access to the IMS data.

The panel will review and evaluate the content and focus of the research support programs of seismology of the Air Force Office of Scientific Research and the Phillips Laboratory of Hanscom Air Force Base.

Soon after the panel began to work on this charge, the Department of Defense announced plans to eliminate the AFOSR and Phillips Laboratory external research program in seismology. The following paragraphs summarize the panel's review of the programs before the study was modified in response to the charge from NTPO.

The AFOSR program used a Broad Agency Announcement (BAA) procedure for soliciting proposals on a broad range of seismological research problems. From 1993-1996 $3.3M/yr of grants/contracts were issued by the 6.1 program with $0.5-0.9M/yr provided to the AFPL to partially support an internal research effort in seismology. The AFOSR program used proposal peer review and relevance reviews by AFTAC and ARPA, with funding decisions based on a combination of value and relevance. The funding for this program was unstable, with annual budget difficulties; however, 57 total grants/contracts were issued. This program bolstered university involvement in CTBT-relevant research, which had diminished significantly as ARPA focused effort on systems development. Research activities conducted by the universities under the AFOSR program included research on regional distance seismograms, elastic and anelastic structure in Eurasia, Africa, the Middle East, and South America, characteristics of Lg propagation, basic wave propagation theory for calculation of regional high frequency phases, source radiation effects in anisotropic media, three component waveform analysis, and numerous other topics relevant to CTBT research. High priority was given to discrimination and location research, moderate priority to magnitude estimation, and relatively low priority to detection and regionalization efforts. Notably, the panel's review of research issues in seismology gives high priority to the last two issues.

The function of transitioning the research developments from the basic (6.1) research program to the operational regime was tasked to the 6.2 program at AFPL. By FY96 the AFPL program involved 8 civil servants and 7 on-site contractors performing directed research efforts. The AFPL budget, mainly for external contracting, was provided by AFOSR, AFPL, AFTAC, ARPA, the Arms Control and Disarmament Agency (ACDA), the State Department and DOE. The FY95 budget for AFPL was $11.3M ($0.72M AFOSR; $1.05M AFPL; $2.74M AFTAC; $2.37M ARPA; $0.6M State Dept.; $3.67M DOE; $0.15M ACDA), and the FY96 budget was $9.74M ($0.9M AFOSR; $0.75 AFPL; $1.74M AFTAC; $1.81 ARPA; $4.40 DOE; $0.15 ACDA). The AFTAC, ARPA, and DOE support administered by AFPL and complemented by grants from the AFOSR 6.1 program, constituted the main research funding support base at universities and some contractors for basic and applied research in CTBT monitoring. The associated total levels were about $12M/yr for FY95 and FY96. Additional support for systems development and advanced developmental research were provided directly by ARPA, AFTAC, and DOE. These funds primarily supported private companies.

In 1996, the AFOSR and Phillips Laboratory (AFPL) programs in CTBT monitoring were eliminated as part of a restructuring of DoD programs in response to changing priorities. While AFTAC continues to be tasked with serving as the CTBT NDC, the AFOSR, AFPL and ARPA research budgets for FY 1997 were consolidated into a single DoD funding line, organized under the new Nuclear Treaty Program Office (NTPO), overseen by the Assistant to the Secretary of Defense for Nuclear, Chemical and Biological Programs.

References

Ad Hoc Committee on a Nuclear Test Ban. 1995. Working Group 1 - Verification International Monitoring System Expert Group, CD/NTB/WP. 224, March 1995.

Ad Hoc Group of Scientific Experts to Consider International Cooperative Measures to Detect and Identify Seismic Events. 1996. Evaluation of the first full year of GSETT-3, GSE/CRP.262. March 29, 1996. 71 pp.

Adushkin, V. V. 1996. Monitoring of underground nuclear tests by seismic stations in the former Soviet Union (FSU). Pp. 35-52 in Monitoring a Comprehensive Test Ban Treaty, E.S. Husebye and A. M. Dainty eds. Dordrecht: Kluwer Academic Publishers.

Bath, M. 1967. Recommendations of the IASPEI committee on magnitudes. Seismology Bulletin. Uppsala, Sweden: Seismology Institute.

Bache, T. C. 1982. Estimating the Yield of Underground Nuclear Explosions. Bull. Seism. Soc. Am., 72(6):131-168.

Bannister, J. R. 1894. Pressure Waves from the Mount St. Helens Eruption, JGR, 89/D3. P. 4895.

Bedard, A. J., Jr. 1977. The D-C pressure summator: Theoretical operations, experimental tests and possible practical uses. Fluidics Quart., 9(1):26-51.

Bedard, A. J., Jr., R. W. Whitaker, G. E. Greene, P. Mutschlecner, R. T. Nishiyama, and M. Davidson. 1992. Proceedings, 10th Symposium on Turbulence and Diffusion, September 29-October 2, 1992, Portland, Ore. Boston, Ma.: American Meteorological Society.

Berkner, L. V., et al. 1959. Report of the Panel on Seismic Improvement. The Need for Fundamental Research in Seismology, Washington, D.C.: Department of State. 212 pp.

Bowyer, T. W., et al. 1996. Automatic radioxenon analyzer for CTBT monitoring, Pacific Northwest National Laboratory- 11424, UC-713.

Buckingham, M. J., and M. A. Garces. 1996. A Canonical Model of Volcano Acoustics., La Jolla, Calif.: Marine Physical Laboratory, Scripps, University of California, San Diego.

Claassen, J. P. 1996. Performance estimates of the CD proposed international seismic monitoring system. Pp. 676-694 in Proceedings of the 18th Annual Seismic Research Symposium on Monitoring A Comprehensive Test Ban Treaty, 4-6 September 1996, J. F. Lewkowicz, J. M. McPhetres, and D. T. Reiter, eds. Hanscom AFB, Ma.: Phillips Laboratory.

Dahlman, O., and H. Israelson. 1977. Monitoring Underground Nuclear Explosions. Holland, Amsterdam: Elsevier Scientific Publishing.

Dainty, A. M. 1996. The influence of seismic scattering on monitoring. Monitoring a Comprehensive Test Ban Treaty, E. S. Husebye and A. M. Dainty, eds. NATO ASI Series. Kluwer Academic Press. 836 pp.

Daniels, F. B. 1959. Noise reducing line microphone for frequencies below 1 Hz.. J. Acoust. Soc. Am., 31:529-531.

Donn, W. L. and N. K. Balachandran. 1981. Mount St. Helens eruption of 18 May 1980: Airwaves and explosive yield. Science, 213:539.

Donn, W., and E. Posmentier. 1967. Infrasonic Waves from the Marine Storm of April 7, 1966. J. Geophysical Res., 72:2053-2061.

Douglas, A. and P. D. Marshall. 1996. Seismic source size and yield for nuclear explosions. Pp. 309-354 in Monitoring a Comprehensive Test Ban Treaty, E. S. Husebye and A. M. Dainty, eds. Dordrecht: Kluwer Academic Publishers.

Dreger, D. S., and D. V. Helmberger. 1993. Determination of source parameters at regional distances with three-component sparse network data. J. Geophys. Res. 98(B5): 8107-8125.

Dziewonski, A. M., T. A. Chou, and J. H. Woodhouse. 1981. Determination of earthquake source parameters from waveform data for studies of global and regional seismicity. J. Geophys. Res., 86:2825-2852.

Einaudi, F., A. J. Bedard Jr., and J. J. Finnigan. 1989. A climatology of gravity waves and other coherent disturbances at the Boulder Atmospheric Observatory during March-April 1884. J. Atmos. Sci., 46(3):303.

Evans, W.C. 1995. Models for assessment of surveillance strategies. Pacific Sierra Research Corporation Technical Note 1079.

Fan, G., S. L. Beck, and T. Wallace. 1994. The seismic source parameters of the 1991 Costa Rica aftershock sequence: Evidence for a transcurrent plate boundary. J. Geophys. Res., 98:15759-15778.

Fan, G., and T. Wallace. 1991. The determination of source parameters for small earthquakes from a single very broadband seismic station. Geophys Res. Lett., 18(8):1385.

Flinn, E. and D. McGowan. 1970. A Least Squares Procedure of Direct Estimation of Azimuth and Velocity of a Propagating Wave. Alexandria Laboratories Report No. AL-70-1. Alexandria, Va.: Geotech.

Glasstone, S., ed. 1957. The effects of nuclear weapons. Dept. of Army Pamphlet No. 39-3.

Green, G.R. and J. Howard. 1975. Natural Infrasound: A One Year Global Study. NOAA Technical Report ERL 317-WPL.

Grover, F. 1971. Experimental Noise Reducers for an Active Microbarograph Array. Geophys. J. R. Astr. Soc., 26:41-52.

Gupta V. 1995. Locating nuclear explosions at the Chinese test site near Lop Nor. Science & Global Security, 5:205-244.

Gutenberg, B. 1945. Amplitudes of surface waves and magnitudes of shallow earthquakes. Bull. Seis. Soc. Am., 35:3-12.

Gutenberg, B., and C. Richter. 1936. Magnitude and energy of Earthquakes. Science (New Series), 83:183-185.

Herrmann, R. B. 1980. Q estimates of the coda of local earthquakes. Bull. Seism. Soc. Am., 70:447-468.

Husebye, E. S., and A. M. Dainty, eds. Monitoring a Comprehensive Test Ban Treaty, Kluwer Academic Publishers, Dordrecht. 836 pp.

Isacks, B., J. Oliver, and L. R. Sykes. 1968. Seismology and the new global tectonics. J. Geophys. Res., 73:5855-5899.

Jabs, R. H., and W. A. Jester. 1976. Development of environmental monitoring system for detection of radioactive gases. Nuc. Tech., 30:24-32.

Jester, W. A., and A. J. Baratta. 1982. Monitoring Krypton-85 during TMI-2 Purging. Nuc. Tech., 56:478-483.

Jester, W. A., A J. Baratta, R. W. Granlund, and G. R. Eidam. 1982. Evaluating radiation monitoring effectiveness for the detection of Krypton-85, Health Physics, 43:827-832.

Johnson, R. E. 1972. An infrasonic pressure disturbance study of two polar substorms. Planet. Space Sci., 20:313-329.

Kushnir, A., V. Lapshin, V. Pinsky, and J. Fyen. 1990. Statistically optimal event detection using small array data, Bull. Seism. Soc. Am., 80:1934-1947.

Lamb, H. 1911. On Atmospheric Oscillations. Proc. Roy. Soc. London, 84A: 551.

Larson, Craine, Thomas, and Wilson. 1971. Correlation of Winds and Geographic Features with Production of Certain Infrasonic Signals in the Atmosphere. Geophysical Journal of R. A. S., 26:201-214.

Levshin, A. L. 1985. Effects of lateral inhomogeneities on surface wave amplitude measurements. Annales Geophysicae, 3:511-518.

Lewkowicz, J. F., J. M. McPhetres, and D. T. Reiter. 1996. Proceedings of the 18th Annual Seismic Research Symposium on Monitoring a Comprehensive Test Ban Treaty, 4-6 September 1996, Phillips Laboratory, PL-TR-96-2153, Hanscom Air Force Base, 1055 pp.

Mack, H., and E. A. Flinn. 1971. Analysis of the Spatial Coherence of Short-Period Acoustic Gravity Waves in the Atmosphere, Geophys. J.R. Astr. Soc., 26:255-269.

Mason, L. R., Jr., and J. B Bohlin. 1995. Optimization of an atmospheric radionuclide monitoring network for verification of the Comprehensive Nuclear Test Ban Treaty. Pacific Sierra Research Corporation Report 2585.

Mayeda, K. 1993. mb(Lg coda): A stable single station estimator of magnitude. Bull. Seism. Soc. Am., 83:851-861.

Meade, C. and D. T. Sandwell. 1996. Synthetic Aperture Radar for Geodesy, Science, 273, 1181-1182.

Murphy, J. R. 1996. Types of seismic events and their source descriptions. Pp. 225-256 in Monitoring Comprehensive Test Ban Treaty, E. S. Husebye and A. M. Dainty, eds. Dordrecht: Kluwer Academic Publishers.

Mutschlecner, J. P. and R. W. Whitaker. 1990. The correction of infrasound signals for upper atmospheric winds. Pp. 143 in Proceeding of the Fourth International Symposium on Long-Range Sound Propagation, William L Willshire, ed. NASA-CP-3101.

Nabelek, J., and G. Xia. 1995. Moment-tensor analysis using regional data: Application to the March 25, 1993, Scotts Mills, Oregon, earthquake. Geophys. Res. Lett., 22(1):13.

National Research Council. 1995. Seismological Research Requirements for a Comprehensive Test-Ban Monitoring System. Washington, D.C: National Academy Press. 80 pp.

Nishiyama, R. T., and A. J. Bedard Jr. 1991. A quaddisc static pressure probe for measurements in adverse atmospheres: With a comparative review of static pressure probe designs. Review of Scientific Instruments, 62:2193-2204.

Nuttli, O. W. 1973. Seismic wave attenuation and magnitude relations for eastern North America. J. Geophys. Res., 78:876-885.

Olson, J.V. 1982. Noise Suppression using data-adaptive polarization filters: Application to infrasonic arrays. J. Acoust Soc. Am., 72(November):5.

Olson, J., C. R. Wilson, J. Collier, and B. McKibben. 1982. Antarctic Atmospheric Infrasound, Final Progress Report, GIR 82-3. Fairbanks, Alaska: Geophysical Institute, University of Alaska.

Olson, J. 1983. Detection and Filtering of Signals in Scalar Arrays, Geophysical Institute Report GIR 83-3 (prepared for AFOSR Nov. 1983). Fairbanks, Alaska: Geophysical Institute, University of Alaska.

Olson, J., C. R. Wilson, and D. Spell. 1983. Annual Report: Antarctic Atmospheric Observations., Report GIR 83-3. Fairbanks, Alaska: Geophysical Institute, University of Alaska.

Patton, H. J. 1980. Surface-wave generation by underground nuclear explosions releasing tectonic strain. Lawrence Livermore National Laboratory, Livermore, Calif., UCRL-53062. 16 pp.

Patton, H. J. 1988. Application of Nuttli's method to estimate yield of Nevada Test Site explosions recorded on Lawrence Livermore National Laboratory's digital seismic system. Bull. Seism. Soc. Am, 78:1759-1772.

Patton, H. J., and W. R. Walter. 1993. Regional movement: Magnitude relations for earthquakes and explosions. Geophys. Res. Let., 20:277-280.

Patton, H. J., and G. Zandt. 1991. Seismic movement tensors of western U.S. earthquakes and implications for the tectonic stress field. J. Geophys. Res., 96:18,245-18,259.

Pearce, R.G. 1996. Seismic source discrimination at teleseismic distances. In Monitoring a Comprehensive Test Ban Treaty, E.S. Husebye and A. M. Dainty eds. Dordrecht: Kluwer Academic Publishers.

Perkins, R. W., and L. A. Casey. 1996. Radioxenons: Their role in monitoring a Comprehensive Test Ban Treaty. Pacific Northwest National Laboratory, DOE/RL-96-51.

Pierce, A. D. and J. W. Posey. 1971. Theory of the Excitation and Propagation of Lamb's Atmospheric Edge Mode from Nuclear Explosions. Geophys. J. R. Astron. Soc., 26:341.

Pisarenko, V., A. Kushnir, and I. Savin. 1987. Statistical adaptive algorithms for estimations of onset moments of seismic phases, Phys. Earth Planet. Interiors, 47:888-900.

Posmentier, E. S. 1967. A Theory of Microbaroms, Geophys. .J. R. Ast. Soc., 13,487.

Pounds, T. J. 1994. A chronology of Comprehensive Test Ban Proposals, Negotiations, and Debates: 1945-1993, Science Applications International Corporation, McLean Virginia.

Richter, C. 1935. An instrumental earthquake magnitude scale. Bull. Seis. Soc. Am., 25:1-32.

Reed, Jack W. 1969. Climatology of Airblast Propagation from Nevada Test Site Nuclear Airbursts. Sandia National Laboratories Report, SC-RR-69-572.

ReVelle, D. O. 1995. Historical Detection of Atmospheric Impacts by Large Bolides. Presentation at the International Conference on Near-Earth Objects, sponsored by Sandia National Laboratories, Explorers Club and the United Nations, April 24-25, 1995.

Ringdal, F. 1985. Study of magnitudes, seismicity, and earthquake detectability using a global network. The VELA Program, a 25 Year Review of Basic Research. Ann Kerr, ed. Executive Graphic Services. 620 pp.

Ringdal, F., and T. Kvaerna. 1989. A multi-channel processing approach to real time network detection, phase association, and threshold monitoring. Bull. Seism. Soc. Am., 79:780-798.

Ritsema, J. and T. Lay. 1993. Rapid source mechanism determination of large Mw > 5 earthquakes in the western United States, Geophys. Res. Lett., 20:1611-1614.

Romney, C. F. 1985. VELA overview: The early years of the seismic research program. Pp. 38-66 in The VELA Program A Twenty-Five Year Review of Basic Research. A. U. Kerr, ed. Defense Advanced Research Projects Agency. Executive Graphic Services.

Sampson, J. and J. Olson. 1981. Data Adaptive Polarization Filters for Multichannel Geophysical Data. Geophysics, 46:1423.

Schlittenhardt, J. 1988. Seismic Yield Estimation Using Teleseismic P- and PKP-waves Recorded at the GRF-(Graefenberg) array. Geophys. J., 95:163-179.

Shearer, P. 1994. Global seismic event detection using a matched filter on long-period seismograms. J. Geophys. Res., 99:13713-13725.

Sipkin, S. A. 1982. Estimation of earthquake source parameters by the inversion of waveform data: Global seismicity, 1981-1983. Bull. Seis. Soc. Am., 76:1515

Song, X. J., D. V. Helmberger, and L. Zhao. 1996. Broadband modeling of regional seismograms; the basin and range crustal structure. Geophys J. Int. 125:15-29.

Stevens, J. L., J. R. Murphy, and N. Rimer. 1991. Seismic source characteristics of cavity decoupled explosions in salt and tuff. Bull. Seism. Soc. Am., 81:1272-1291.

Stevens, J. L., W. I. Rodi, J. Wang, B. Shkoller, E. J. Halds, B. F. Mason, and J. B. Minster. 1982. Surface wave analysis package and Shagan River to SRO station path corrections. S-CUBED Topical Report (VSC-TR-82-21 SSS-R-82-5518), La Jolla, Calif.

Stewart, S.W. 1977. Real-time detection and location of local seismic events in central California. Bull. Seism. Soc. Am., 67:4333-4452.

Sykes, L. R. 1996. Dealing with decoupled nuclear explosions under a comprehensive test ban treaty. Pp.247-293 in Monitoring a Comprehensive Test Ban Treaty, E.S. Husebye and A. M. Dainty, eds. Dordrecht Kluwer Academic Publishers.

Taylor, S. R., M. D. Denny, E. S. Vergino, and R. E. Glaser. 1989. Regional discrimination between NTS explosions and western U.S. Earthquakes. Bull. Seismol. Soc. Am., 79:1142-1176.

Thio, H. K., and H. Kanamori. 1995. Moment Tensor Inversions for local earthquakes using surface waves recorded at TERRAscope. Bull. Seism. Soc. Am., 85(4):1021.

Thurber, C. H., H. R. Quin, and P. G. Richards. 1993. Accurate locations of nuclear explosions in Balapan, Kazakhstan, 1987 to 1989. Geophys. Res. Lett., 20:399-402.

U.S. Department of Energy. 1993. CTBT Technical Issues Handbook. UCRL-ID-117293.

U.S. Department of Energy. 1994. Comprehensive Test-Ban Treaty Research and Development FY95-96 Program Plan. Department of Energy, DOE/NN-0003.

United States of America. 1994. Working Paper, Monitoring a comprehensive test ban treaty: An overview of the U.S. Approach, CD/NTB/WP.53, Conference on Disarmament, 19 May. 9 pp.

von Seggern, D. and R. Blandford. 1972. Source time functions and spectra for underground nuclear explosions, Geophys. J., 31:83-97.

Wallace, T. C., and D. V. Helmberger. 1982. Determination of seismic parameters of moderate-size earthquake from regional waveforms. Phys. E-Plan., 30:185-196.

Walter, W. R. 1993. Source parameters of the June 29, 1992 Little Skull Mountain earthquake from complete regional waveforms at a single station. Geophys Res. Lett., 20(5):403.

Whitaker, R. W., J. P. Mutschlecner, M. B Davidson,. and S. D. Noel. 1990. Infrasonic observations of large-scale HE events. Pp. 133 in Proceedings of the Fourth International Symposium on Long-Range Sound Propagation, W. L Willshire, ed. NASA-CP-3101.

Wilson, C. R. 1969a. Auroral infrasonic waves. J. Geophys. Res., 74:1812-1836.

Wilson, C.R. 1969b. Two-station auroral infrasonic wave observations. Planet. and Space Sci., 17:1817.

Wilson, C. R. 1969c. Infrasonic waves from moving auroral electrojets. J. of Planet. and Space Physics, 17:1107.

Wilson, C. R. 1971. Auroral infrasonic waves and pole-ward expansions of auroral substorms at Inuvik, N.W.T., Canada. Geophysical Journal of R.A.S., 26:179-181.

Wilson, C. R. 1974. Trans-auroral zone auroral infrasonic wave observations. Planet. Space Sci., 22:151-173.

Wilson, C.R., J. Olson, and R. Richards. 1996. Library of Typical Infrasonic Signals (LTIS). ENSCO Subcontract No. 269343-2360.009. Geophysical Institute, University of Alaska, Fairbanks, Alaska.

Wimer, N. G. 1997. Capt. USAF, AFTAC, Personal communication with W. A. Jester.

Woodhouse, J. H. 1974. Surface waves in a laterally varying layered structure. Geophys. J. R. astr. Soc., 97:461-490.

Woods, B. B., and D. G. Harkrider. 1995. Determining surface-wave magnitudes from regional Nevada Test Site data. Geophys. J. Int., 120(2):474

Woods, B. B., S. Kedar, and D. V. Helmberger. 1993. M_L:M_0 as a regional seismic discriminant. Bull. Seism. Soc. of Amer., 83(4) 1167-1183.

Young, C., J. Beiriger, J. Harris, S. Moore, J. Trujillo, M. Withers, and R. Aster. 1996. The waveform correlation detection system project: Issues in system refinement, tuning and operation. Pp. 789-798 in The Proceedings of the 18th Annual Seismic Research Symposium on Monitoring a Comprehensive Test Ban Treaty, 4-6 September, 1996, J. Lewkowicz, J. McPhetres, and D. Reiter, eds.

Young, J. M., and W. A. Hoyle. 1975. Computer program for multidimensional spectra array processing. NOAA TR ERL 345-WPL. Boulder, Col.: NOAA, Environmental Research Laboratories.

Zhao, L. S., and D. V. Helmberger. 1994. Source estimation from broadband regional seismograms. Bull. Seismol. Soc. Am. 84:91-104.

Zhu, L. and D. V. Helmberger. 1996. Advancement in source estimation techniques using broadband regional seismograms, Bull. Seism. Soc. Am., 86:1634-1641.

A

Statement of Task

Charge from the Nuclear Treaty Program Office

The panel will review and evaluate the following:

1) What are the basic research problems remaining in the fields of seismology, hydroacoustics, infrasonics and radionuclides that should be pursued to meet national and international requirements for nuclear monitoring? The panel's work on this question should anticipate quality of data to be made available in the future, in particular those data from the CTBT International Monitoring System.
2) What research is necessary to strengthen the synergy between the seismic, hydroacoustic, infrasonic, and radionuclide data sets to improve overall monitoring capability and to meet national and international requirements?
3) How should the research results be transitioned so that they are most useful to those responsible for monitoring and verifying a CTBT?
4) What are characteristics of a long-term program that would provide a stable, but adaptable base of support to those responsible for monitoring and verifying a CTBT?

Original AFOSR-Phillips Laboratory Seismic Review Panel

The panel will review and evaluate the content and focus of the research support programs of seismology of the Air Force Office of Scientific Research and the Phillips Laboratory of Hanscom Air Force Base.

B

Research Support History

On September 2, 1959, the Department of Defense (DoD) initiated a research program to improve national capabilities to detect, identify, and determine characteristics of foreign nuclear explosions in response to the 1959 Berkner panel report, with funding made available to the Advanced Research Projects Agency (ARPA). A few weeks later ARPA order Number 102 was signed, funding the U.S. Air Force for research in nuclear test monitoring technologies. This initiated the VELA program, which had the goals of lowering the detection threshold, developing effective identification criteria, and developing On-Site Inspection techniques with the associated goal of improving the seismic location accuracy (Romney, 1985). Efforts to achieve those goals, driven to ever more demanding levels by ensuing nuclear testing treaties, have continued to the present, although the formal involvement of ARPA in the research effort terminated with a DoD restructuring in 1996.

There were substantial research and development programs in multiple disciplines from 1959 on, including deployment of VELA satellites and research on On-Site Inspection, tunneling technology, and other areas. In recent years, CTBT-related budgets have primarily supported seismic and satellite monitoring. Since satellite monitoring is outside the scope of the panel's report, this appendix will focus on the support for seismic monitoring. Prior to 1970 there were large expenses involved in the deployment and operation of seismic arrays (e.g., the Large Aperture Seismic Array (LASA), VELA arrays, NORSAR, WWSSN), and the cumulative spending on seismology during this period was $202 million (not corrected for inflation). For ensuing years, the ARPA seismological fundamental research and seismic array development budgets (apart from operations at AFTAC) are given in Table B.1

Up until 1983, ARPA-funded portions of the budgets that went to universities were at levels of around $6 million per year. The Air Force provided AFOSR with about $0.5 million per year of additional funds. From 1983 to 1990, ARPA funding for fundamental research was administered by the predecessor to the Air Force Phillips Laboratory (AFPL), with an increasing percentage of the ARPA budget being used to develop regional arrays and establish computer facilities at the Center for Seismic Studies (CSS). Ultimately, the CSS evolved into the prototype-IDC.

The distribution of funding among basic, applied, and systems development activities has varied during the past few years. In 1990, Congress withdrew funds from ARPA in order to

TABLE B.1 ARPA Seismic Research Budgets (not corrected for inflation)

Year	(Funding $million)	Year	(Funding $million)
1970	$31	1979	$8
1971	$20	1980	$11
1972	$19	1981	$15
1973	$12	1982	$18
1974	$10	1983	$18
1975	$9	1984	$14
1976	$12	1985	$14
1977	$11	1986	$15
1978	$9	1987	$16

establish the Air Force 6.1 program administered by AFOSR for "a university based research program." This program provided funding for university seismological research from 1991 to 1996. During this period the majority of the remaining ARPA budget was focused on development of the prototype IDC, including computer and communications systems development, software automation, and operational software development.

The AFOSR program used a Broad Agency Announcement (BAA) procedure for soliciting proposals on a range of seismological research problems. From 1993 to 1996 $3.3 million per year of grants or contracts were issued by the 6.1 program with $0.5 to 0.9 million per year provided to the AFPL to partially support an internal research effort in seismology. The AFOSR program used proposal peer review and relevance reviews by AFTAC and ARPA, with funding decisions based on a combination of value and relevance. Funding for this program was unstable, with annual budget difficulties, and many leading seismologists chose not to participate; however, 57 grants or contracts were issued. This program bolstered university involvement in CTBT-relevant research, which had diminished significantly as ARPA focused effort on systems development. Research activities conducted by universities under the AFOSR program included research on regional distance seismograms; elastic and anelastic structure in Eurasia, Africa, the Middle East, and South America; characteristics of Lg propagation; basic wave propagation theory for calculation of regional high-frequency phases; source radiation effects in anisotropic media; three-component waveform analysis; and numerous other topics relevant to CTBT research. High priority was given to discrimination and location research, moderate priority to magnitude estimation, and relatively low priority to detection and regionalization efforts.

The function of transitioning research developments from the basic (6.1) research program to the operational regime was tasked to the 6.2 program at AFPL. By FY 1996 the AFPL program involved eight civil servants and seven on-site contractors performing directed research efforts. The AFPL budget, mainly for external contracting, was provided by AFOSR, AFPL, AFTAC, ARPA, the Arms Control and Disarmament Agency (ACDA), the State Department, and the Department of Energy (DOE). The FY 1995 budget for AFPL was $11.3 million ($0.72 million, AFOSR; $1.05 million, AFPL; $2.74 million, AFTAC; $2.37 million, ARPA; $0.6 million State Department.; $3.67 million, DOE; $0.15 million, ACDA), and the FY 1996 budget was $9.74 million ($0.9 million, AFOSR; $0.75 million, AFPL; $1.74 million, AFTAC; $1.81 million, ARPA; $4.40 million, DOE; $0.15 million, ACDA). The AFTAC, ARPA, and DOE support administered by AFPL and complemented by grants from the AFOSR 6.1 program constituted the main research funding support base for universities and some contractors for basic and applied research in CTBT monitoring. The associated total levels were about $12 million per year for FY 1995 and FY 1996. Additional support for systems development and advanced developmental research were provided directly by ARPA, AFTAC, and DOE. These funds primarily supported private companies.

The DOE external research program, initiated in 1995 with internal funding, provided the FY

1995 and FY 1996 DOE funds administered by AFPL. DOE provided $3.665 million in FY 1995 and $4.395 million in FY 1996 for this program. (This external support has almost been eliminated in the DOE FY 1997 budget.) The two-year research program supported by DOE and other AFPL sources was broader in scope than the AFOSR 6.1 program and included research on seismic, hydroacoustic, infrasonic, and radionuclide monitoring technologies. (A description of the full scope of DOE's internal and external program during this period is given in DOE, 1994.) The AFPL program spanned a range of basic and applied research efforts and had a BAA solicitation along with peer review of the proposals.

The DOE CTBT research program has a FY 1997 budget of $69.6 million. This budget involves $45.1 million for satellite systems development, $11.6 million for research on seismic methods, $1.2 million for research on hydroacoustic methods, $1.1 million for infrasonic methods, $3.4 million for radionuclide systems, $6.5 million for automatic data processing development, $0.6 million for statistics, and $0.2 million for On-Site Inspection development, distributed across four National Laboratories.

Beginning with FY 1997, the entire AFOSR, AFPL, and ARPA budgets for CTBT monitoring were consolidated under the newly created Nuclear Treaty Program Office (NTPO) with a FY 1997 budget of $29.1 million, with $8.8 million intended for a peer-reviewed external CTBT monitoring research program ($7.1 million specifically for research in seismology and $1.7 million for research in other disciplines such as hydroacoustics, infrasonics, and radionuclides). The $20.3 million balance was for sustained systems development for the IDC. The NTPO has indicated that the external research program will be administered by the Defense Special Weapons Agency (DSWA). Given the reduction of DOE external funding for CTBT research occurring in the FY 1997 budget, the $8.8 million support level represents a major decrease of the overall funding support for basic and fundamental research for the expanded set of CTBT monitoring disciplines. The President's FY 1998 budget includes a continuation of DSWA-sponsored research at the $8.8 million level. DOE commitments to the external CTBT research program continue to decrease.

C

Seismic Event Location

Event location is an essential procedure in CTBT monitoring, playing a critical role in characterizing and identifying every source. Operational considerations related to On-Site Inspections have established a goal that remote treaty monitoring methods routinely locate sources on land to within an area of 1000 km^2 or less. In practice, it is common for the areal uncertainty associated with seismic locations to be much greater than this, even for events located by using large numbers of stations. This appendix explains why the problem exists and suggests some ways that location uncertainties can be reduced to 1000 km^2 level. However, no single method of improved analysis will lead to the necessary improvement on a global basis. Universally improved event location procedures require a systems approach and calibration effort.

These efforts could, in turn, greatly benefit all users of global seismic data. Traditionally, data used to estimate the origin time and location (latitude, longitude, and depth) of an earthquake or an explosion are the arrival times of various seismic waves measured at stations situated around the world. If seismic arrays are available, it is also possible to measure the directions from which the seismic waves arrive at the array. To a limited extent this can also be done with three-component stations. Such data are then interpreted by using a model of the Earth's velocity structure (i.e., a description of the velocity of seismic waves throughout the Earth's interior or travel time curves). By starting with a trial location (e.g., beneath the station that reports the earliest arrival time) and origin time, and calculating the travel time from the source to the station based on the distance and the velocity model, arrival times can be calculated at each station. These can be compared with the actual arrival time, and by iteratively revising the origin time and location to improve the match between measured and calculated arrival times, a solution can be found that gives the smallest difference between the observed arrival times and the times predicted for that Earth model.

Examination of the way in which computed arrival times change for perturbations in locations in the vicinity of the "best-fitting" location determines a relationship between the random uncertainty in measured arrival times and the size of the region in which the source is expected to lie. Such uncertainty is conventionally reported in terms of a "90 per cent confidence error ellipse," a type of two-dimensional confidence interval that would contain the actual solution 90 times out of 100 if there were no systematic error (as discussed below). If the region of uncertainty were circular, the area corresponding to the CTBT location accuracy goal of less than 1000 km^2 would have a radius of 17.84 km. The size and shape of the error ellipse depends on the random uncertainties in arrival time measurements, the number and geographic distribution of the stations that record the arrivals, and (unknown) errors in the velocity model of the Earth. In practice, it is desirable to have detections from stations within at least two azimuthal quadrants from the event (and preferably from three or from all four) to reduce the triangulation errors incurred in working back from the detecting stations to the source location.

Since the random error in measuring the arrival time of seismic waves is usually less than 1 second (generally less than 0.1 second when signal-to-noise ratios are good) and since the velocity of seismic waves is typically less than 6 km/s in the Earth's outer layers where the events of interest occur and where measurements are made, it might appear that seismic sources can routinely be located to within a few kilometers, with corresponding areal uncertainty of only a few tens of square kilometers. However, this conclusion is incorrect at present because the lack of a sufficiently good model of the Earth's velocity structure introduces systematic errors or biases, sometimes called model errors. These errors are the principal problem in determining locations and in estimating the associated location uncertainty.

At depths greater than about 200 km, the Earth's global velocity structure is known quite accurately (i.e., to within about ±1 per cent, except in regions of subducting tectonic plates where the variability can be greater). At shallower depths, however, and within the crust in particular (which varies in thickness from 5 to 75 km), the velocities of seismic waves may differ from the velocities in a given seismic model in unknown ways by ±5 per cent, or even more in some regions. These are not random uncertainties but reflect a fundamental lack of information about the material properties and conditions in these regions. As a consequence, the arrival times of teleseismic waves are affected in ways that are not accounted for by standard simple Earth models (which are usually assumed to be spherically symmetric velocity distributions), and this in turn can result in systematic mislocation of the sources in a given region.

The situation is even more complex for locations determined with data from regional seismic stations. The arrival times of regional waves depend strongly on the extremely heterogeneous, shallow crustal structure. Earth models often have a uniform crustal layer, and the deviations between the actual and calculated arrival times are often larger than found for teleseismic observations. As a result, event locations based only on regional arrival times or small numbers of teleseismic and regional arrivals are often poor. It is a common experience when locating a moderate or large earthquake with many teleseismic stations that inclusion of regional observations and use of a simple regional crustal model actually degrade the event location (unless the stations are close to the source so that little time is spent in the anomalous region and significant travel time errors do not accumulate).

Seismic arrays and three-component stations can provide constraints on the source back-azimuth in addition to providing arrival time information, and this additional information clearly assists in the process of triangulating on the source. However, estimates of back-azimuth are also vulnerable to misinterpretation because of uncertainties in the Earth's velocity structure, unless corrections are made. The effect of inadequately modeled Earth structure for direction of approach at the array is somewhat different than in the case of interpreting arrival time data, but the result is still that a location can be quite poor and the associated estimates of location uncertainty may be wrong (i.e., the true source location may be inside the 90 per cent confidence ellipse estimated using the erroneous Earth model far less often than 90 per cent of the time). Location uncertainty estimates are made using an intrinsically inaccurate model of the Earth, and the effects of (unknown) systematic deviations between the Earth and the model are hard to quantify and to include in the source uncertainty estimate. In seismological practices there are some efforts to estimate the model uncertainty by statistical approaches, comparisons of results for different reference models, and direct measurement using events with known locations. Without such efforts, event location estimates must be viewed with skepticism.

There are three principal ways to work around the problem of ignorance of Earth structure: (1) use numerous stations at different azimuths and different distances around the source in an attempt to average out the differences between the Earth's actual velocity structure and that of the model; (2)

improve the information about the Earth's velocity structure and thus determine a more sophisticated and presumably more accurate model, that includes variability; and (3) empirically "calibrate" the station (or array) so that, in effect, the source of interest is located with reference to another event with an accurately known location near the event of interest. In this approach, data for the unlocated event are usually "corrected" for "path anomalies" determined from observations of calibration events at each station, and the corrected arrival times are used to locate the event by using a standard Earth model. In some cases, the differences in the arrival times of the unlocated and calibration events are calculated directly and used to obtain a "relative location" between the reference event and the event of interest. Such relative locations can have much higher precision than raw locations, but the accuracy of the absolute locations will only be as good as those of the calibration events, at best.

The USGS/NEIS, a number of large regional and national networks, and the International Seismological Centre (ISC) all use strategy 1 for routine processing of event bulletins. The global or regional Earth models tend to be simple one-dimensional models that predict arrival times simply as a function of distance from the source. The general approach to improving location accuracy has been to add stations. Changes in the reference model are resisted because of a desire for uniformity of the historical catalog and because, with large numbers of observations, the locations are not strongly dependent on the details of one-dimensional models. Larger events that are recorded by larger numbers of stations tend to have smaller location uncertainties, whereas small event locations tend to have larger uncertainties due to the decreased resolution, the relatively larger effects of heterogeneity of paths, and the greater potential for bias associated with small numbers of observations. Regional networks with hundreds of stations separated by tens of kilometers have been deployed in seismically active areas where accurate location of small events (even down to magnitude 1 or smaller) is deemed of importance. The performance of these networks hinges on the proximity and number of the nearest stations to the sources.

The research community often uses all of these strategies to study special sets of events and to develop three-dimensional velocity models. In many cases it appears that improved locations are obtained, but earthquake monitoring operations have been slow to embrace laterally varying Earth models or station corrections. There is not extensive operational experience with methods 2 and 3 on a global basis, but it is clear that method 1 can achieve global location accuracies at the 1000 km^2 level only for quite large events recorded by large numbers of stations. CTBT monitoring will involve many small events recorded by small numbers of stations, even when IMS and NTM are combined, so some form or combination of strategies 2 and 3 is imperative, and no clear alternatives exist.

The above approaches can result in greatly improved locations, but none of them can do a reliable job of characterizing the uncertainty of the final location estimate until accurate three-dimensional Earth models become available. Often, the difference between actual and assumed Earth structure results in locations estimates that, for a particular region, are all shifted in the same direction, perhaps a few tens of kilometers from the true locations. The closer a three-dimensional model approaches a description of the true Earth, the better will be the estimates of location uncertainty made with that model.

There is a long history of coming to grips with systematic error in seismic estimates of explosion locations. Early United States experience with nuclear explosions in Nevada was used to develop a model of the Earth's crust in that region, and when the first underground explosion in the United States outside Nevada was carried out in New Mexico in 1961, it was estimated by assuming that New Mexico had a Nevada-type crust to have a depth of 130 km! (An event with such a depth estimate would normally be identified with high confidence as an earthquake, unless the formal uncertainty estimate on depth was comparable. More generally, interpretation of the event location is one of the simplest and most widely used discriminants, which again is a reason for working to obtain the best possible location estimates.) Even in areas that have been studied care fully with calibrated stations over a period of many years, ground truth has shown that seismic locations were not as good as had been thought. For example, for the last 20 explosions at the Balapan region of the Semipalatinsk test site, Thurber et al. (1993) showed that locations determined to within about ±100 m from SPOT satellite photographs were outside the seismically determined 95 per cent confidence ellipses in most cases. In this case, the ellipses were only about 5 to 10 km^2 in area and had been determined by using the known location of a reference event. The ellipses would have included the actual locations 95 per cent of the time if they had been enlarged to about 20 km^2, so in this case, seismic locations were actually quite good. Yet the fundamental problem remains—until some type of ground truth becomes available, the size of the confidence ellipses does not account for model inaccuracies. Thus, there are great uncertainties in translating knowledge of Earth structure into errors for event locations.

Although the use of large numbers of stations can reduce the location error, Figure C.1 shows that the area of an error ellipse decreases with event size down to a certain amount, but then does not get much smaller even for large explosions (when hundreds of stations contribute arrival times). Uncertainties in Earth structure limit the value of additional data.

From the standpoint of solving the seismic event location problem in the U.S. CTBT monitoring context, the ultimate seismological solution is to work toward an improved three-dimensional model of the velocity structure for the regions of interest to the United States, since this will give the most direct interpretation of monitoring data. Models of the Earth will always be simplified because its total complexity is unknowable. Experience has shown, however, that sufficiently complex models can be constructed for regions of interest so that locations are accurately known and sufficiently precise for applications such as CTBT monitoring or analysis of earthquake faulting. However, the goal of developing regional models (or a global model) with such detail is an undertaking of much greater effort than the usual research project. A small group of individuals working for a year or two is not going to solve this problem. What is needed is a systems approach.

There are about 20 earthquakes per day at magnitude 4 and above, whose signals can be used to interpret global and regional Earth structure. Events of smaller magnitude can also be used to learn about regional structure if nearby stations are available. The crux of the problem, however, is that the locations of these events are not known independently when trying to improve the model of Earth structure. At best, for the vast majority of events, locations will be estimated based on models that are only approximations to structure. It, therefore, appears that location parameters must be determined at the same time as the parameters of the velocity model. Many researchers have explored ways to carry out such simultaneous determinations, which can be made to work on the scale of a local network as well as on a global scale. It has also been demonstrated that complete modeling of the full set of regional waveforms can improve constraints on the source depth and epicenter and provide information about the crustal structure that is difficult to extract from arrival times alone.

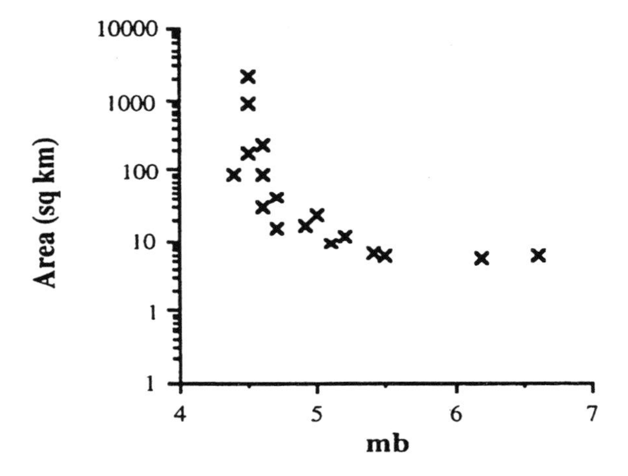

FIGURE C.1 Variation of event location 95 per cent confidence ellipses as a function of m_b for events at the Chinese test site with calibration by a satellite location for one event. SOURCE: Gupta, 1995.

Systematic efforts to determine details of regional and global structure are being conducted, funded by earthquake monitoring agencies, CTBT research programs, and basic Earth science programs, but there is no concerted effort to integrate these into a global model. The community interested in CTBT monitoring could undertake a long-term program to develop a sufficiently accurate three-dimensional Earth model, with laterally varying crustal and lithospheric structure, so that all events at and above the monitoring threshold could be located with the desired precision. Such an effort would be coordinated with the earthquake monitoring and basic Earth science research communities because these groups would benefit from systematically improved event locations at local and global levels.

An alternative strategy for making progress is an empirical approach 3 of calibrating stations and arrays by building up an archive of events whose location is known accurately. Calibration efforts could be performed on relatively localized regions, which can be tuned to specific U.S. CTBT monitoring priorities.

The prototype-IDC has begun to set up such a calibration event data base. Although this prototype IDC effort is a step in the right direction, it is thought of in terms of only a few events per day. A more ambitious approach could be taken to build up an archive of accurate event locations from much larger sets of available data and use these improved locations to calibrate stations on a much more extensive scale in areas of interest, not just stations used by the IMS and NTM. To get accurate locations for the purposes of station calibration, it is possible to use locally recorded large mine blasts and earthquakes whose location becomes well known as a result of rupture of the ground surface, reports of strong ground shaking, or data provided by a good local network or a mining company. This empirical approach would result in a steady cycle of improvement: better locations can lead to better calibration of new stations and better knowledge of Earth structure, which in turn leads back to better locations.

To address the immediate problem that a treaty monitoring network has in locating a new event quickly, the key is to maintain as large an archive as reasonably possible of accurately located seismic events and of their signals at the network of stations used for monitoring the new events. Comparison of the new signals with the old can then lead to a location estimate that starts with the old event and adds the relative location of the new event. This result typically can be better than an estimate made directly from the arriving signals without any comparison to events in the archive.

To address the long-term problem of how to build up the archive by continuing to add well-located events, a commitment is needed to develop a comprehensive bulletin of seismicity down to low magnitudes in areas of interest (using teleseismic and regional signals from large numbers of stations) that emphasizes accuracy of location, rather than speed of production.

D

Seismic Magnitudes and Source Strengths

Assessment of the seismic monitoring of a CTBT requires working definitions of the threshold level in terms of seismic magnitude, and several discriminants depend on accurate determination of the seismic magnitude (in one or more frequency ranges). For example, one of the most successful and proven teleseismic methods to discriminate seismic events is the classical m_b: M_S comparison, for which explosions have a significantly smaller surface wave magnitude (M_S, measured from the peak long-period surface wave amplitude at a period of about 20 seconds) than an earthquake with the same body wave magnitude (m_b, measured from the peak short-period P-wave amplitude, typically near a period of 0.5-1.0 seconds. The effectiveness of this discriminant is attributable to the difference in the characteristic dimensions of the two sources: an earthquake ruptures over a plane which is large in size relative to the cavity created by an underground explosion. One can also think of this source difference in terms of the time function excitation, which tends to be long for earthquakes and short and impulsive for explosions. Important contributions to this discriminant also come from the inefficient excitation of Rayleigh waves from an isotropic source and the rebound of the explosion cavity which produces a peak in the amplitude spectrum.

Significant challenges have arisen as the monitoring community has attempted to extend this and related teleseismic discrimination methods to smaller events recorded at regional distances: the relative source dimensions change, fewer stations record signals from the events, the paths are more variable and complex, and the complexity introduces differences in numbers and characteristics of the signals (e.g., frequency content and amplitude). These differences require definitions of source magnitude that are (1) consistent with the teleseismic estimates for large events that are recorded both regionally and teleseismically; (2) are applicable in those cases where only regional stations detect signals; and (3) can be normalized from one region to the next. This appendix reviews the nature of seismic magnitudes along with more physically well-defined measures of source strength involving seismic moment and seismic wave energy derived from waveform modeling approaches.

LOCAL MAGNITUDE, M_L

The procedure for assigning magnitudes to seismic sources originated with Richter (1935), who used recordings made on a specific instrument (the Wood-Anderson) to estimate the relative size of local earthquakes in southern California. He constructed an empirical standard curve that characterized the variation in observed amplitudes (log of the peak amplitude on any component) with distance from an event. This empirical curve was then used to reduce the measurement made at the actual epicentral distance of the seismometer to that expected for a seismometer at 100 km. This magnitude scale is known as M_L and is localized to southern California. Figure D.1 provides examples of the differences in signals from nuclear explosions and earthquakes at the Nevada Test Site (NTS) recorded at a broadband station in Pasadena, California (the signals are filtered to correspond to various classical instruments). The local magnitudes for these two events, averaged over the array of simulated Wood-Andersons from broadband stations in southern California, are $M_L = 5.6$ for the explosion (Kearsarge: the U.S. joint verification experiment [JVE] explosion) and 5.7 for the Yucca Mountain earthquake. Note that there is an order-of-magnitude larger surface wave for the earthquake compared to the explosion despite their comparable short period amplitudes that result in similar local magnitudes.

BODY WAVE MAGNITUDE, m_b

One of the most common measures of seismic source strength (m_b) is based on the P-wave amplitude. Essentially, a peak-to-peak amplitude (*A*) measurement is taken from the first 4 seconds of each record of a short-period vertical component recording or array beam along with an estimate of the period (*T*) of the peak motion. After *A* is corrected for the instrument amplification factor at that period, the body wave magnitude is calculated from the average of measurements at j stations: $m_b^j = \log_{10}(A_j/T_j) + B(D_j) + S_j$ ($j = 1, \ldots, j$). Here $B(D_j)$ is an empirical correction factor for source depth and epicentral distance [conceptually, m_b can be calculated for stations at any distance, as long as the $B(D_j)$ value is established for the particular phase being measured, such as Pn, P, or PKP], and S_j is an empirical station correction factor based on relative station amplitude patterns. Averaging observations is essential, given the large variance in magnitude measurements for a given event (for measurements without station corrections this varies over one full magnitude unit or more, as a result of heterogeneous propagation effects or azimuthal variations in the radiation patterns, and even when station corrections are used, there is a typical scatter of at least 0.3 unit of magnitude). Statistical methods can be used to allow for amplitude measurements that are off-scale or that are below the noise to give the most consistent relative magnitude determinations. As events of interest become smaller, the number of observations that contribute to their magnitude estimation decreases, and the associated uncertainty in magnitude value increases. The development of threshold estimates or discrimination techniques for small events must account for these factors.

Because explosions can be idealized as rapidly occurring isotropic pressure sources, they often produce strong impulsive, relatively simple P waves at teleseismic distances. Thus, one of the earliest methods for estimating explosion yields was based on relating m_b to yield empirically at teleseismic distances. Following the standard m_b versus yield approach (e.g., Dahlman and Israelson, 1977), known events with known yields are used to construct a curve (usually a straight line) relating m_b to yield. When m_b is measured for a new event, the curve is used to estimate the yield corresponding to that m_b.

Given the close cooperation between Russian and U.S. seismologists in past decades, yields are now known at the major test sites, and curves of m_b versus yield are well established for medium size underground explosions. Adushkin (1996) gives m_b (ISC) = 0.86 log W + 3.87 for the Nevada Test Site, and m_b (ISC) = 0.77 log W + 4.45 for the East Kazakhstan test site for well-coupled events with yields less than 150 kt. Note the large difference in m_b for 1 kt yield events: 3.87 versus 4.45. This results from a combination of source coupling and upper-mantle attenuation effects for each test site. The m_b (ISC) values are determined without any station corrections, but rely on relatively large numbers of stations, so data censoring effects are not likely to be too severe except for the largest and smallest events used to determine the above relationships.

Events with $m_b < 4$ are usually recorded only regionally, and in Adushkin's study these were calibrated against larger events for the same paths by assuming that the entire NTS area is a uniform source region. Thus, the B(D) corrections to m_b,

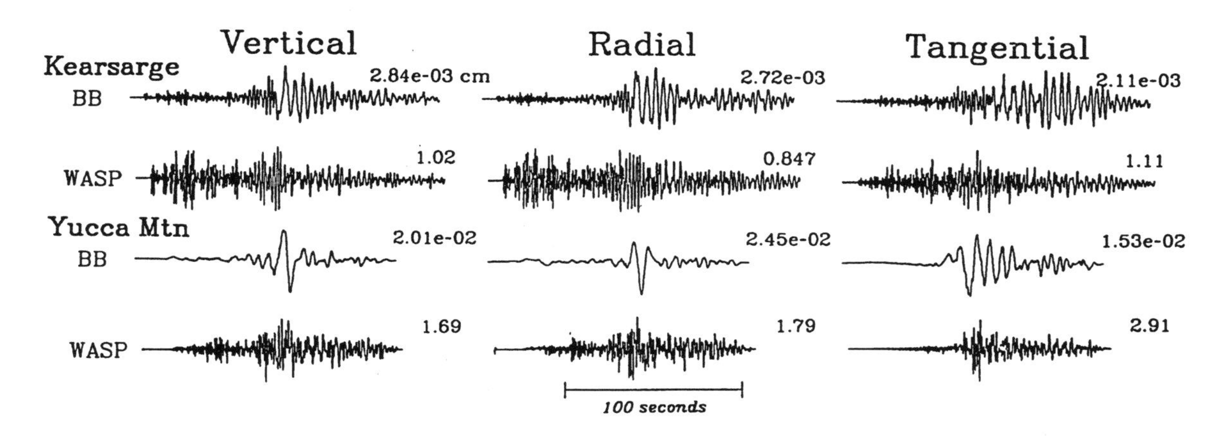

FIGURE D.1 Three-component broadband records for the Kearsarge Joint Verification Experiment at the Nevada Test Site and a Yucca Mountain earthquake near NTS recorded in Pasadena, California. Simulated short-period records for the events (derived from broadband measurements) are indicated as WASP. The amplitudes of the displacements for all records are noted in centimeters. Although there is an order-of-magnitude difference in the amplitude of the broadband data, M_L for the earthquake and the explosion are comparable (5.6 Kearsarge, 5.7 Yucca Mountain.) because of their similar short-period response (WASP). Source: D. Helmberger.

which are poorly constrained at regional distances, become less important since the station corrections for this well-calibrated source area absorb any offset. However, as pointed out by Adushkin (1996), the scatter in M_L (Berkeley) relative to NEIS m_b was -0.2 to +0.7 units for the 14 smallest shots considered, and this high variance is not explained. In general, measurements of m_b using regional phases have high variance.

SURFACE WAVE MAGNITUDE, M_S

A magnitude scale based on teleseismic surface waves was described by Gutenberg and Richter (1936) and developed more extensively by Gutenberg (1945). It used the amplitude of maximum horizontal ground displacement due to surface waves with periods around 20 seconds and is called M_S. This scale was generalized by Bath (1967) to

$$M_S = \log (A/T) + 1.66 \log D,$$

where A is the peak-to-peak amplitude of the Rayleigh wave (amplitude measured in millimicrons on the vertical component) and D is epicentral distance in degrees.

The U.S. Geological Survey (USGS) NEIS uses this formula, but restricts measurement to the period range $18 < T < 22$ seconds, and thus explicitly avoids regional Rayleigh waves (Airy phase) such as those displayed in Figure D.1, which have shorter predominant periods. The requirement of measurable 20-second-period Rayleigh wave arrivals is quite demanding for small events, particularly for isolated seismometers. As a result, conventional M_S is effectively restricted to events with $m_b > 4.0$-4.5. However, since the Rayleigh waves at Pasadena (PAS) for the NTS Kearsarge explosion look quite similar to those for lower-yield events (Woods et al., 1995), it appears viable to estimate M_S at all ranges if a stable correction for the signal period is made. This proves difficult because the strength of the regional Rayleigh waves depends on many factors: source type, source depth, local structure, and attenuation.

The approach taken for extending M_S to regional distances by Woods and Harkrider (1995) involves modeling the Rayleigh waves produced by NTS explosions at all distances by assuming a mixed path correction, essentially a local structure at NTS and a regional structure along the path to the station obtained from prior efforts of Stevens et al. (1982). The formalism of

Levshin (1985) is used, which treats the propagation of Rayleigh waves along a slowly varying inhomogeneous path as developed earlier by Woodhouse (1974). These approximations are used routinely in the Harvard Centroid-Moment Tensor (CMT) solutions. The Woods and Harkrider (1995) study determined an effective long-period seismic moment M_0 first and then determines an equivalent teleseismic M_S. They report on 50 North American stations observing 190 NTS explosions where the regional events are on-scale for small events and not observed at some of the more distant stations. They obtain $M_S = \log M_0 + b$, where b = 11.38 for NTS.

A similar study by Stevens for the Semipalatinsk, East Kazakhstan, test site involving primarily teleseismic data obtained b = 11.86. Thus, NTS appears less effective in producing surface wave amplitudes per unit of source strength M_0 or yield than does the Soviet test site. However, the difference in frequency content between explosions and earthquakes implied by the East Kazakhstan experiment suggests that m_b:M_S discriminants should be even more effective in central Eurasia than in the western United States.

Lg MAGNITUDE

Given the difficulties of estimating m_b from Pn for small events and recognizing that Lg signals are typically larger than Pn signals, Nuttli (1973) introduced the Lg magnitude scale m_b (Lg). The short-period Lg phase defined by Nuttli has an apparent velocity of about 3.5 km/s. The phase generally looks slightly dispersed in character as seen in Figure D.1 (Lg arrives about a minute after the first arrivals [Pn] in these observations). Nuttli assumed the Lg arrival to be formed from the superposition of higher-mode surface waves, and because superposition combines energy leaving the source at many different angles, it should require fewer measurements at fewer stations than m_b to suppress radiation pattern effects (Dainty, 1996). Nuttli picked the third largest amplitude in the window formed by velocities 3.2 to 3.6 km/s as representative of strength and reduced the measurement to a distance of 10 km similar to Richter's local magnitude approach. He defined $m_b(Lg) = 5 + \log_{10} [A(10 \text{ km})/110]$, where A is measured in millimeters and corrected for attenuation by assuming a decay of A proportional to $\exp[-\alpha(D \times 10)]$ (see Douglas and Marshall, 1996, for a discussion).

The attenuation parameter α is estimated from the actual record directly, as discussed by Herrman (1980). Nuttli (1985) showed that NTS yields can be estimated remarkably well by applying this methodology to a few regional stations. The procedure has been repeated by Patton (1988) using some changes in definitions and yielded similar results. Thus, if a path can be calibrated well, this measure proves highly stable compared to m_b (Pn). Unfortunately, Lg is subject to path blockage caused by complexity in crustal structure in some regions.

CODA MAGNITUDE

Stable single-station estimates of magnitude made using the envelope of the 1 Hz Lg coda show promise for regional seismic monitoring (Mayeda, 1993). The Lg coda is generated by scattering, but decay of the Lg coda appears to be controlled by anelastic losses. Amplification effects near the recording stations and attenuation control the absolute amplitude, but the decay rate of the coda depends on the average medium properties of the crust sampled by the coda waves. Given limited calibration information from a region, coda magnitudes appear to provide high-precision estimates of magnitude using the data recorded at even one station.

The amplitude of the Lg coda is modeled by the equation:

$$A_c(t) = N\, t^{-a} \exp(-bt),$$

where $A_c(t)$ is the noise-corrected, time-dependent Lg coda amplitude; N contains both source and receiver effects; t is time in seconds, and a and b are constants representative of the medium from the path to the receiver (a and b are determined by fitting the shape of the curve to the noise-corrected amplitudes for the Lg coda). Given these values, the coda magnitudes are normalized to $m_b(Lg)$ by assuming that $\log_{10}$ (N = $m_b(Lg)$ + constant, where the constant provides the station correction. Once the station correction is determined, the equation can be used to determine an estimate of $m_b(Lg)$ for a new event, given the measurement of $A_c(t)$ for that event.

SOURCE STRENGTH ESTIMATION BASED ON WAVEFORM MODELING

The explosion monitoring methodology in current use has developed around the magnitude scales described above. Some efforts were conducted by ARPA during the mid-1980s to place yield estimation on a modeling basis, but uncertainties in how to treat the surface interaction (pP) in the presence of spall, near-field attenuation, tectonic release, and so forth, proved too large to be competitive with the more direct empirical procedures (see Murphy, 1996). Large events, which provide many observations for averaging, allow meaningful comparison of explosion and earthquake sources, but near the monitoring threshold there will be few observations of each event. The assumption of averaging over many stations to remove source and path effects begins to break down when extensive averaging cannot be performed. Operational questions arise for smaller events: for example, should the station magnitude corrections determined by processing teleseismic signals be applied to observations at closer distances; should magnitude corrections be derived by systematically correcting for biasing effects of earthquake source orientation as suggested by Pearce (1996)?

An alternative to the empirical approach of estimating source strengths involves waveform modeling methodologies. Examples of routine earthquake processing with such approaches include Harvard's solutions for long-period body and surface waves (Dziewonski et al., 1981) and USGS moment tensors from teleseismic P waves (Sipkin, 1982). The former method uses a three-dimensional Earth model and locates the event as it finds the optimal source representation. The latter method relies on the NEIS location. Both approaches can provide globally complete focal mechanism solutions for events above m_b = 5.0-5.5, but various strategies allow similar routine source inversions to be made for smaller events using regional signals (e.g., Ritsema and Lay, 1993).

SOURCE MECHANISM ESTIMATION FROM REGIONAL SEISMOGRAMS

New analytical tools have been developed recently to estimate source parameters from broadband regional seismic data. Two kinds of regional waveform data are typically used for source estimation: surface waves (e.g., Patton, 1980; Patton and Zandt, 1991; Thio and Kanamori, 1995) and body waves (e.g. Wallace and Helmberger, 1982; Fan and Wallace, 1991; Fan et al., 1994; Dreger and Helmberger, 1993). Generally, body waves are less affected by shallow heterogeneities and are more stable than surface waves, although they have a lower signal-to-noise ratio due to their lower energy. There have been several inversion methods proposed recently using whole seismograms (Ritsema and Lay, 1993; Walter, 1993; Zhao and Helmberger, 1994; Nabelek and Xia, 1995). Most of these inversions are controlled mainly by surface waves, particularly because they are performed using long-period waveforms.

An exception is the "cut-and-paste" (CAP) method of Zhao and Helmberger (1994), which breaks broadband waveforms into *Pnl* and surface wave segments and inverts them independently. The source mechanism is obtained by applying a direct grid search through all possible solutions to find the global minimum of misfit between the observations and synthetics, allowing time shifts between portions of seismograms and synthetics to reduce the influence of poorly known structure. One advantage of the technique is that it proves insensitive to velocity models and lateral crustal variation. Source resolvability as a function of stations and components is discussed in Zhao and Helmberger (1994). Song et al. (1996) performed a detailed sensitivity study and showed that a single two-layered crustal model produces an adequate set of synthetics to fit most earthquakes in the western United States if time shifts are allowed. More detailed modeling may provide better fits and less shifting but has little effect on the resulting source parameter estimates. The inversion discussed above has been further stabilized by including absolute amplitudes and is fully automated on the TERRAscope data stream in southern California (Zhu and Helmberger, 1996).

MAGNITUDE DIFFERENCES BETWEEN EXPLOSIONS AND EARTHQUAKES

Various comparisons of long-period surface wave magnitude (M_S or M_0) levels can be made with short-period body waves [m_b, m_b(Pn), M_L] to assess characteristic spectral behavior of different source types. Of particular importance is the behavior seen at regional distances. A plot of M_L versus M_0 for a population of NTS explosions and western U.S. earthquakes is given in Figure D.2. The type of source was assumed to be known in determining M_0 values, so this is not a discrimination plot, but an indication of physical differences between the sources.

There is a significant separation of earthquake and explosions using this criterion, with no real overlap of the two populations although several earthquakes plot near the explosion region. This separation exists at all sizes. It appears that explosions and earthquakes follow separate respective scaling laws over a wide range of local magnitudes and moments. For earthquakes there is relatively simple scaling over seven and a half orders of magnitude. Earthquakes with a log moment below 13.0 nm were determined from local stations (D < 1°) and would not be detectable at larger distances. They are included here to show the continuity of the linear scaling relationship between M_L and log M_0 for earthquakes.

There is significant scatter in both data sets. The amount of stress relieved by an earthquake is a function of the material properties of the source region and the shape of the surface over which the fault ruptures. Higher-stress drop events are richer in high-frequency energy; thus, the ratio of short- to long-period magnitude fluctuates somewhat from one earthquake to the next. The spectral signature of explosions is also known to depend on the emplacement medium. Explosions can also release ambient stress within the Earth's crust. This tectonic release creates a secondary seismic source that can modify the energy radiation pattern as well as the spectral character of the source. These effects can vary the short- to long-period magnitude ratio of explosions.

These differences of source spectra are quite similar to those presented by Patton and Walter (1993) for the well-calibrated region around the

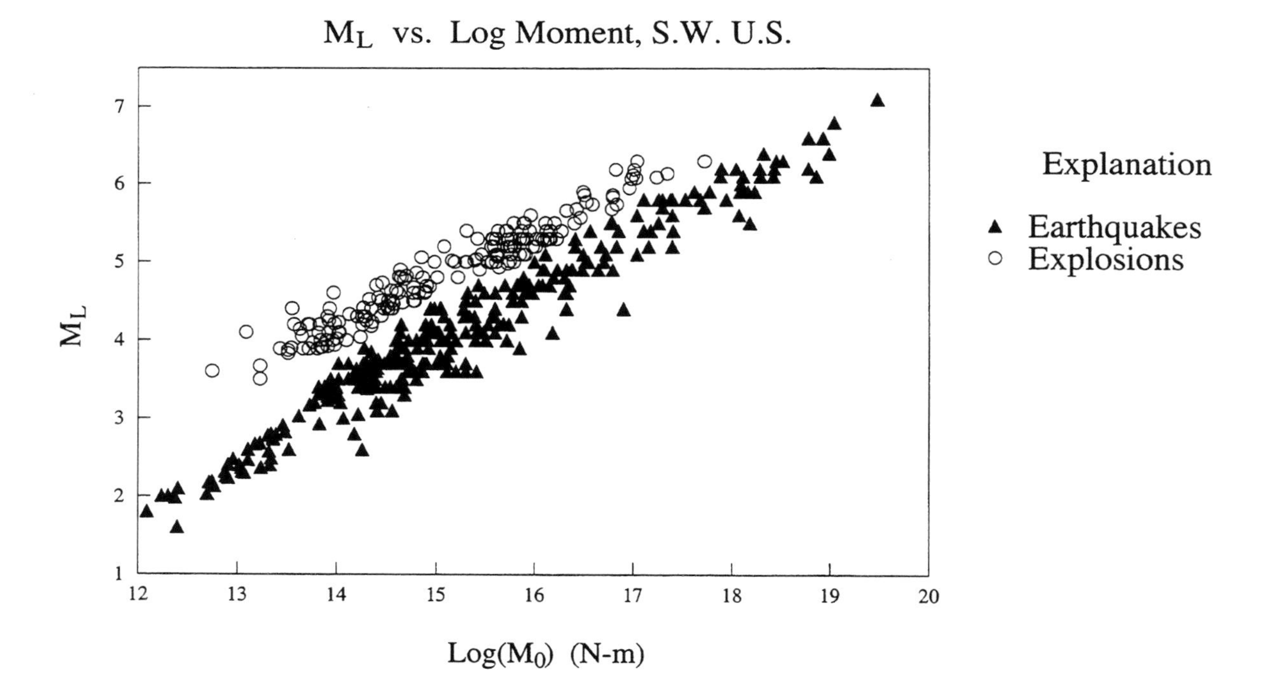

FIGURE D.2 Local magnitude versus event seismic moment (M_0) for earthquakes and explosions in the Southwestern United States. Source: D. Helmberger.

NTS in terms of $m_b(Pn):M_0$ and $m_b(Lg):M_0$ and the results presented earlier by Taylor et al. (1989) on regional $m_b(m_{7.9}):M_S$. Some regional discriminants are designed to detect these source differences.

ROLE OF SIZE ESTIMATION IN TREATY MONITORING

The initial stages of underground explosion monitoring were driven almost entirely by empirical relationships involving these magnitudes. It was noted 30 years ago that P waves from underground shots become quite simple at distance ranges beyond 30° (3300 km) and that the existing m_b (teleseismic body wave magnitude as introduced by Beno Gutenberg) was a convenient indicator of source strength. Consequently, m_b-yield relationships have been used to estimate yields globally with a long and ultimately successful history (Schlittenhardt, 1988) involving an instructive interplay between policy and technology. Similarly, surface wave magnitudes at teleseismic distances M_S were used to establish empirical M_S-yield relationships. For well-coupled explosions in a tectonically active area such as NTS, the empirical relationship

$$m_b = 4.05 + 0.75 \log Y,$$

where m_b is determined teleseismically from a large number of stations and W is yield in kilotons, proves quite effective (Murphy, 1996). Offsets in the regression line became obvious when comparing NTS data to those for explosions at the Amchitka test site (e.g., von Seggern and Blandford, 1972). Distinct differences in plots of m_b versus M_S for U.S. and Soviet tests were also easily recognized (e.g., Bache, 1982). These features were resolved after considerable scientific and policy debate by allowing for differences in upper-mantle attenuation levels. For example, yield estimates based on m_b for well-coupled underground explosions at the Semipalatinsk test site are well matched by

$$m_b = 4.45 + 0.75 \log Y.$$

The m_b values are adjusted upward relative to NTS because of lower attenuation in the mantle under this site (e.g., Murphy, 1996). The two relationships given above were derived before their validation by JVE tests, which is considered to be one of the triumphs of the U.S. research and monitoring community.

Theoretical source models that have been developed to match observed seismic signals, such as the Mueller-Murphy model, prove quite effective at predicting teleseismic P-wave amplitudes in diverse media. For example, the m_b generated by events in clay- or water-filled cavities can be 0.5 magnitude units higher (factor of 3 greater amplitudes) than for normal hard rock sites, whereas a reduction of 0.5 unit is expected for events in dry porous media (Murphy, 1996).

Although spherical source models for explosions work well for explaining P waves, they do poorly in predicting the remaining portions of seismograms, especially at regional distances where measures of ML and Lg are made. The difficulty is that these observations are usually controlled by shear waves (see Figure D.1), which in principle are not excited by an explosion source. Shear wave excitation by explosions is partly explained by triggered tectonic release, but this is difficult to predict, especially in remote regions with no testing history. Tectonic release does appear to produce scatter in M_S for some test sites, which has made mb the more attractive measure of yield.

Much of the source strength research conducted by the seismic monitoring community has addressed events with yields greater than 100 kt relevant to the Threshold Test Ban Treaty (TTBT). These events have m_b values near 6 and are well recorded globally, typically with more than 30 stations reporting magnitude values at teleseismic distances where the paths travel deep in the Earth and avoid upper-mantle complexity. Although individual distance-corrected P-wave amplitudes at these ranges can vary by factors of 10, the network average magnitude tends to be quite stable when large numbers of observations are available. Station corrections and source region biases established empirically for large events can be applied to smaller events located near the larger ones as long as signals are detectable at the same stations. For yields of 1 kt, which have seismic magnitudes near 4.0, only a few teleseismic observations are usually available. This requires use of regional data to estimate mb. At regional distances, amplitude-distance curves become highly variable, and they are unknown in the regions where many of the new IMS stations are being installed.

At regional distances it is possible to use empirical magnitude measures and to define station corrections similar to those at teleseismic ranges. However, regional distance signals have more complex propagation effects, and as a result, source strength estimation may be better performed by quantitative modeling procedures that synthesize complete seismic signals, explicitly accounting for propagation complexities, or use features such as the coda of the wave, which are less dependent on specific paths. Modern instrumentation, coupled with recent modeling techniques, allow complete long-period (>30-second) regional signals to be modeled down to magnitude 4.5-5.0 with simple velocity models (Ritsema and Lay, 1993), whereas shorter-period data (down to 5-second period) can be modeled for events as small as magnitude 3.5 in regions with well-determined velocity structures such as California (Zhu and Helmberger, 1996).

E

Hydroacoustics

This appendix is intended to provide the nonexpert reader with a basic understanding of sound propagation in the ocean.

THE SOFAR CHANNEL

The ocean is a remarkably efficient medium for sound propagation, due largely to the existence of the deep sound or SOFAR (SOund Fixing and Ranging) channel. The channel is characterized by a minimum in the vertical sound speed profile, which occurs at about 1.5 km depth near the equator and gradually rises to shallower depths as one progresses in either direction toward the poles, until the minimum reaches the surface in the Arctic and Antarctic Oceans. Sound speed increases with increasing temperature and increasing pressure (density). The sound speed minimum is a result of higher sound speed at the surface where waters are warm, gradually diminishing with depth as temperature decreases, and then beginning to increase again due to the competing effect of increasing pressure. In northern regions there is little surface heating so sound speed simply decreases with depth due to increasing pressure.

The channel forms a natural waveguide, and sound energy is refracted toward the axis of the channel and away from surface and bottom boundaries. (Figure E.1 shows a sound velocity profile [SVP] and a ray trace.) As a result, sound energy that is coupled into the channel is not attenuated by scattering that would occur if it were to strike the surface or bottom, and the only losses result from geometric spreading of the wavefield and absorption due to conversion to heat arising from viscous and ionic relaxation effects. At low frequencies, absorption losses are small, and geometric spreading accounts for most reduction in signal level.

In shallow waters or at distances from a deep water source comparable to the water depth, the wavefront expands cylindrically, and sound pressure geometric loss is proportional to $1/r$. In practice, transmission losses are usually calculated more precisely by a variety of numerical solutions of the scalar wave equation.

As a consequence of the sound channel and the primarily $1/r$ spreading loss, relatively low-energy signals can be detected at long range. For example, 1 kg TNT explosions at SOFAR axial depths are detected easily at distances of several thousand-kilometer range. Earthquakes and seismic prospecting signals from explosives and airguns can be heard well above the background noise at such ranges.

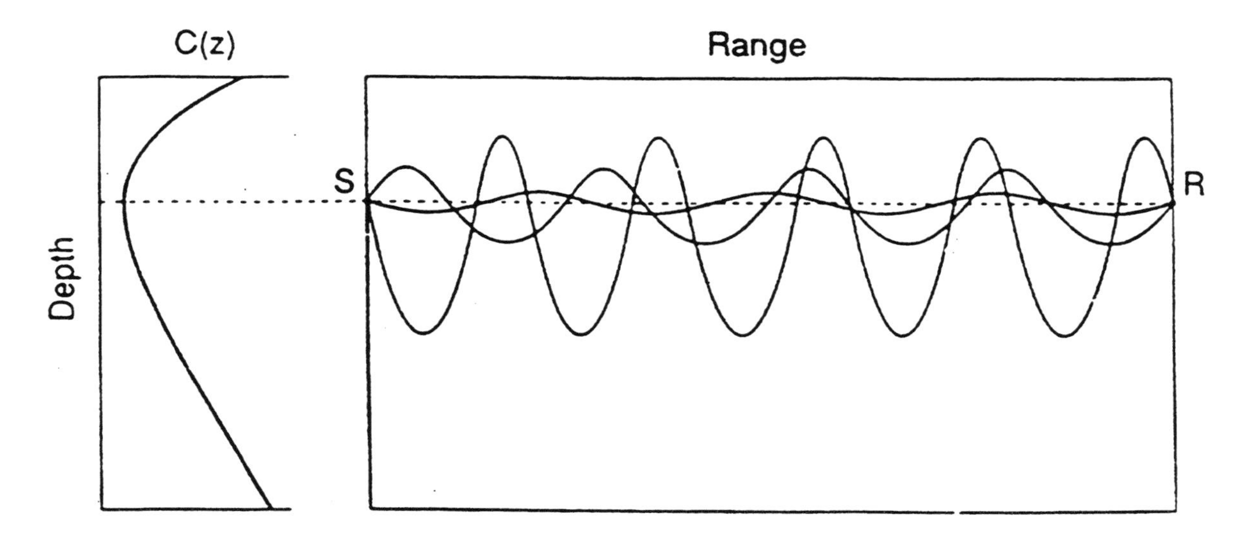

FIGURE E.1 Velocity profile (left) and ray diagram (right) for an idealized SOFAR channel showing ray paths to 700 km distance.

AMBIENT NOISE

There is a persistent level of background noise in the ocean that arises as a result of both man-made and natural processes. At the lowest frequencies (<<10 Hz) the sources of noise include seismicity, ocean turbulence, volcanism, the non-linear interaction of gravity waves (swell), and shallow water gravity waves. In the spectral region between 10 and 100 Hz the noise field is dominated by worldwide ocean shipping and by wind-driven breaking waves. These higher-frequency sources create a more or less continuous red noise spectrum, high at the lowest frequencies and decreasing at 8 to 10 dB per octave up to about 20 Hz, where the spectrum flattens somewhat due to the influence of shipping noise. Rising above this spectral background are distinct noises such as vocalizations by marine animals (whale sounds can be as loud as ships), lightning striking the sea surface (which generates locally sharp sound impulses), and sounds generated by oil and gas prospecting and drilling operations. Hydrocarbon exploration often involves the use of explosive charges or explosive-like signals generated by airgun arrays.

The noise field produces two main monitoring challenges. First, signals from nuclear explosions must be detected against the background noise in which they are embedded. Second, nuclear explosions must be distinguished from noise impulses due to other types of explosions or explosive-like signals.

UNDERWATER EXPLOSIVE SOURCES

Underwater chemical explosions have been studied intensively since World War II, and their characteristics are well understood and accurately modeled. These form a major source of information for monitoring underwater nuclear explosions because of the lack of data from such explosions.

When an explosion occurs at depth, the pressure in the water nearby is so great that the wave velocity becomes a function of pressure and a steep-wavefront, nonlinear shock wave is developed. The shock wave travels radially outward, gradually diminishing in amplitude and entering the linear propagation regime where the wave velocity is constant. For a 1 kt underwater nuclear explosion, the transition from nonlinear to linear propagation occurs at about 10 km.

Hot gases resulting from the explosion are contained by hydrostatic pressure within a bubble, which expands rapidly. As the bubble expands, the pressure inside decreases. The momentum of the velocity becomes a function of pressure and a steep-wavefront, nonlinear shock wave is developed. The shock wave travels radially outward, gradually diminishing in amplitude and entering the linear propagation regime where the wave velocity is constant. For a 1 kt underwater nuclear explosion, the transition from nonlinear to linear propagation occurs at about 10 km.

Hot gases resulting from the explosion are contained by hydrostatic pressure within a bubble, which expands rapidly. As the bubble expands, the pressure inside decreases. The momentum of the water continues the expansion of the bubble beyond the point at which the internal pressure falls below the external hydrostatic pressure, and the bubble contracts, thereby compressing the gas until its pressure is sufficient to halt the motion of the water, whereupon the cycle repeats, each time with diminished intensity. The oscillating bubble generates a series of pressure pulses, called bubble pulses, which are characteristic of deep underwater explosions. Under ideal conditions it is possible to observe numerous bubble pulse oscillations. Note, however, that long range transmission losses at low frequencies are variable and they can have a major impact on the potential of the bubble pulse as a discriminant. If the explosion is shallow, the bubble vents directly into the atmosphere, and no bubble pulse signature is observed. If the explosion is located above the surface the amount of sound energy coupled into the water is orders of magnitude less than an underwater explosion and again there is no bubble pulse. A 1 kt explosion well coupled to the sound channel (detonated, for example, at 1000 m depth) generates a sound pressure level between 300-310 dB relative to 1 μPa at 1 m (depending upon the depth) and has a bubble pulse period of 0.7 second. A modern airgun array used for seismic exploration can have a sound pressure level of 264-270 dB relative to 1 μPa at 1 m (depending upon volume and pressure of the airgun array) and no discernible bubble pulse.

INTERNATIONAL MONITORING SYSTEM HYDROACOUSTIC SIGNAL LEVELS

A signal-to-noise ratio of about 10 dB is usually required to ensure robust detection. With a source level of 280 dB (1 kiloton [kt] explosion at depth), a background 10 Hz noise field of 80 dB (heavy shipping), and a signal integration time of only a few seconds, this signal will remain 50 dB or more above the background noise at global ranges and can be detected easily and distinguished from other sources. However, there are regions in which it may be blocked by bathymetry or attenuated by scattering that arises, for example, from upward refraction in a shoaling sound channel in Arctic waters and scattering from the sea surface and bottom.

A 1 kt explosion detonated 1 km above the sea surface has a calculated source level at the sound channel axis that is 35 dB less, comparable to the intensity of an airgun array. Its intensity at an IMS hydroacoustic sensor can easily be less than the background noise or comparable in level to earthquakes and airguns.

F

Infrasonics

This appendix provides some additional technical details about infrasonics theory and practice, along with identifying some multiuse applications of the International Monitoring System (IMS) infrasonic data.

INTRODUCTION

Low-frequency acoustic signals in the Earth's atmosphere are traditionally termed infrasound when they have frequencies below 10 Hz. Although subaudible, they do propagate as regular acoustic signals. However, because of the low frequency, there is little physical absorption of signal energy. There is little excess loss beyond geometric spreading, which is always present.

Infrasonic systems must typically detect pressure changes levels one millionth of the total atmospheric pressure in the midst of pressure changes from a variety of sources. These "contaminating" sources include high winds, turbulence, eddies, thermal plumes, atmospheric gravity waves, and Earth vibrations. For example, normal wind speeds from 5 to 15 m/s can generate quasi-static dynamic pressures from two to three orders of magnitude greater than most infrasonic signals.

Atmospheric acoustic or gravity waves are lower-frequency infrasonic waves indicating that dual processes can influence atmospheric propagation. For normal sound waves the restoring force controlling propagation is the compressibility of the medium. At longer periods, gravitational restoring forces start to become important. Hence, the term acoustic or gravity wave is sometimes used for lower-frequency infrasound. Figure F.1 indicates the approximate frequency range at which this transition takes place.

Lamb waves are atmospheric surface waves that travel along the Earth's surface at the speed of sound in air and can be compared to Stoneley waves in seismology. They represent solutions to the atmospheric equations that have a totally horizontal wave vector, attenuate exponentially with altitude and with depth into the Earth, and attenuate slowly with range (Lamb, 1911; Pierce and Posey, 1971). In the earlier monitoring period, with rather large source yields, Lamb waves were quite distinct with a long-period cycle arriving before the main acoustic signal (see Figure F.2). For CTBT monitoring the source size of interest is smaller, around 1 kiloton (kt). Based on the analysis of Pierce and Posey (1971), Lamb waves for smaller sources may not be as robust as they are

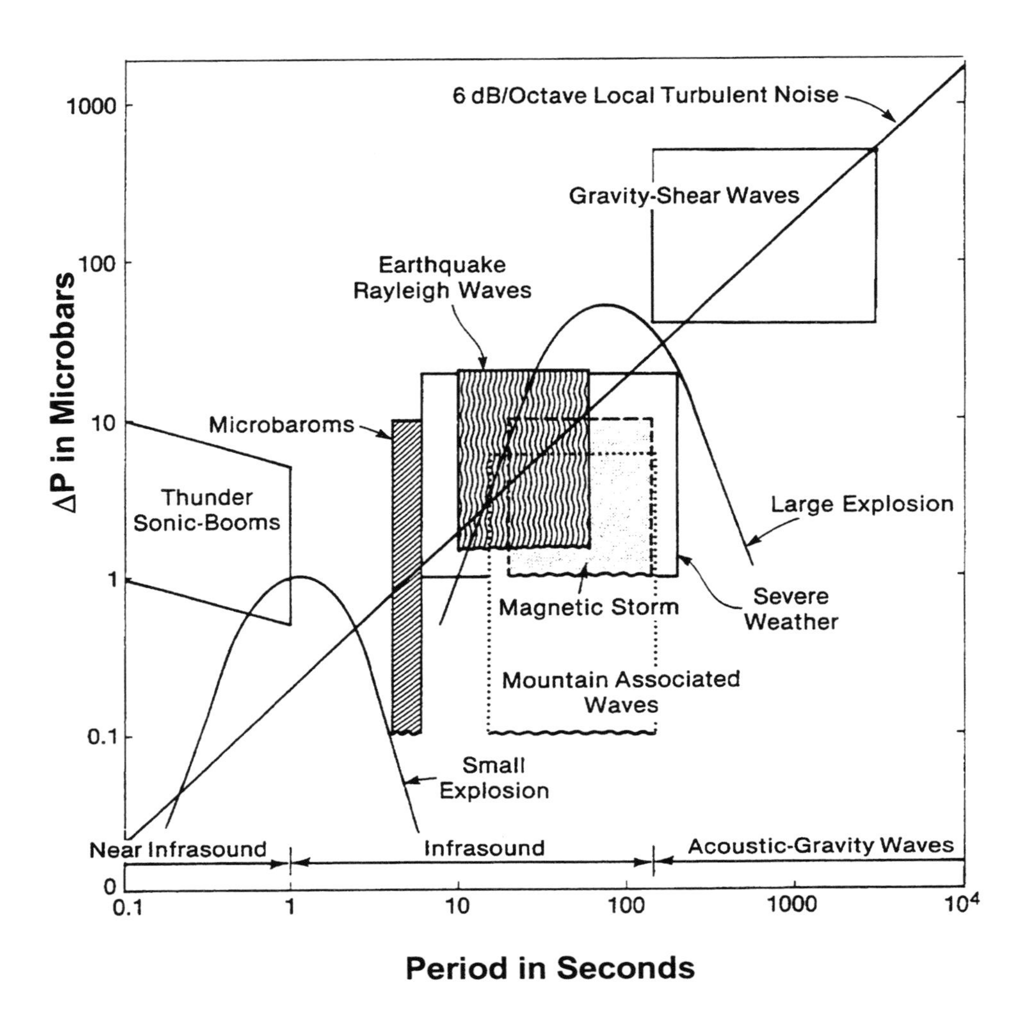

FIGURE F.1 Frequency ranges and signal pressures of infrasonic and acoustic or gravity waves.

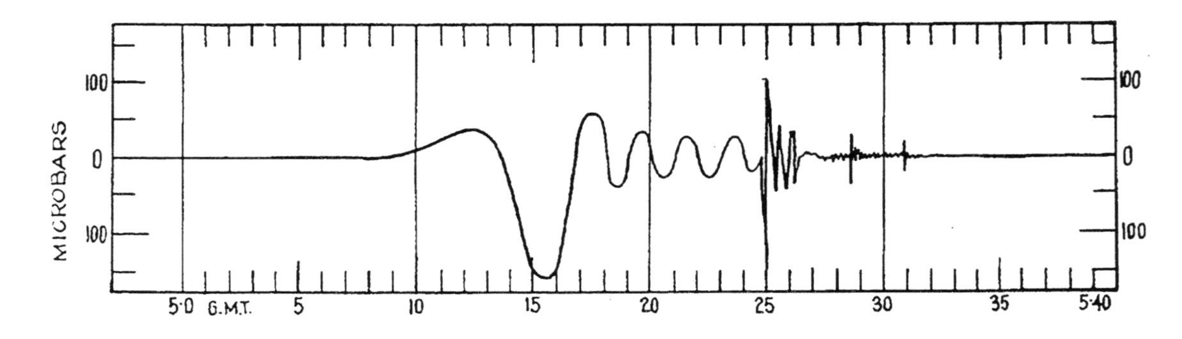

FIGURE F.2 Sample infrasound record of an atmospheric explosion showing a low-frequency Lamb wave arriving before the main acoustic signal. The timescale is indicated in minutes.

for larger sources. The period will approach that of the main acoustic signal and the amplitudes will be less than that of the main acoustic signal. In addition, turbulence and viscosity in the boundary layer can limit their propagation.

SENSOR DESIGN

The unwanted fluctuations can largely be eliminated by sampling over an area and not merely at a point, thus averaging out disturbances with small length scales. Since these "filters" use the spatial distribution to help remove unwanted quasi-static pressure noise (also called pseudosound, because at a point it mimics infrasound), they are known as spatial filters, infrasonic filters, or infrasonic noise reducers.

Figure F.1 summarizes a number of the concepts introduced above and also indicates the approximate frequency ranges for some phenomena producing infrasonic and local pressure signals. It is clear that geophysical signals can appear in the same passband as explosive signals. Figure F.3 is an overview of the wavelengths of various signal types as a function of period; it emphasizes the differences in wavelengths of different sources of pressure fluctuations, with Rayleigh wave-coupled pressure signals having large wavelengths and pressure signals associated with winds and turbulence having small scale sizes. The dimensions of some spatial filters are indicated on the ordinate. (Figures F.1 and F.3 are intended for summary overviews and treat infrasonic monitoring factors in a simplified way.)

Past efforts at creating a single sensor capable of covering a large area (e.g., in the form of large pancake-like devices) have not proven practical nor have large numbers of individual sensors. Pneumatic sampling has been effective. In this approach, pressure signals enter ports or porous openings, pass down pipes or tubes, and are then summed together at the sensor analogous to the summing junction of an operational amplifier. A great range of geometries has been tested, the most widely used types being long pipes with periodic sensing ports at intervals, covering a 300 m distance. More recently, using pipes or porous garden hose, 8 or 16 m radial tubes from a central sensor have allowed cost-effective areal sampling. Such filters greatly reduce unwanted noise, but there is still a need to distinguish true infrasonic signals from the remaining local noise, which does not propagate at acoustic frequencies. Beam-steering processing helps to accomplish this in an effective way.

COMPREHENSIVE NUCLEAR TEST BAN TREATY INFRASOUND MONITORING

If a source produces infrasonic energy, there is a good chance that energy can be detected by remote sensors as is the case for hydrophones in the marine environment. Of particular interest are atmospheric explosions from a fraction of a kiloton to a few kilotons in size. The low-frequency components of the near-field blast wave, as modified by propagation, become the long-range propagating infrasound signal. Experience shows that for 1 kt explosions, detection ranges can be between 2000 and 5000 km depending on noise level and time of year.

Measurements are made with arrays of sensors, usually four or more, so that traditional beamforming techniques can be applied to determine the bearing to observed signals. Thus, an important aspect of detection is the use of correlation among sensor elements in the array. Single-sensor measurements are generally not useful because one cannot tell if increased power is from local noise or from genuine distant sources. Also, the use of power detections alone is not as beneficial as correlation due to the nature of low-level winds, which provide much of the background noise. Bearing estimates are normal outputs of infrasonic array processing and are used to aid in source location. Of course, bearing accuracy is dependent on array size, signal-to-noise ratio, and frequency.

In CTBT applications, sources of interest must be detected by at least two arrays. Single-station events generally will not be passed on for further examination, and this serves as a filter for uninteresting events. With two stations and the same event, the two bearings should intersect and provide a preliminary location.

Noise at an array can come from ground-level winds for which noise-reducing hoses are effective, although for the CTBT bandpass, work needs to be done to define the optimum noise- reducing configuration. There can also be "noise" from

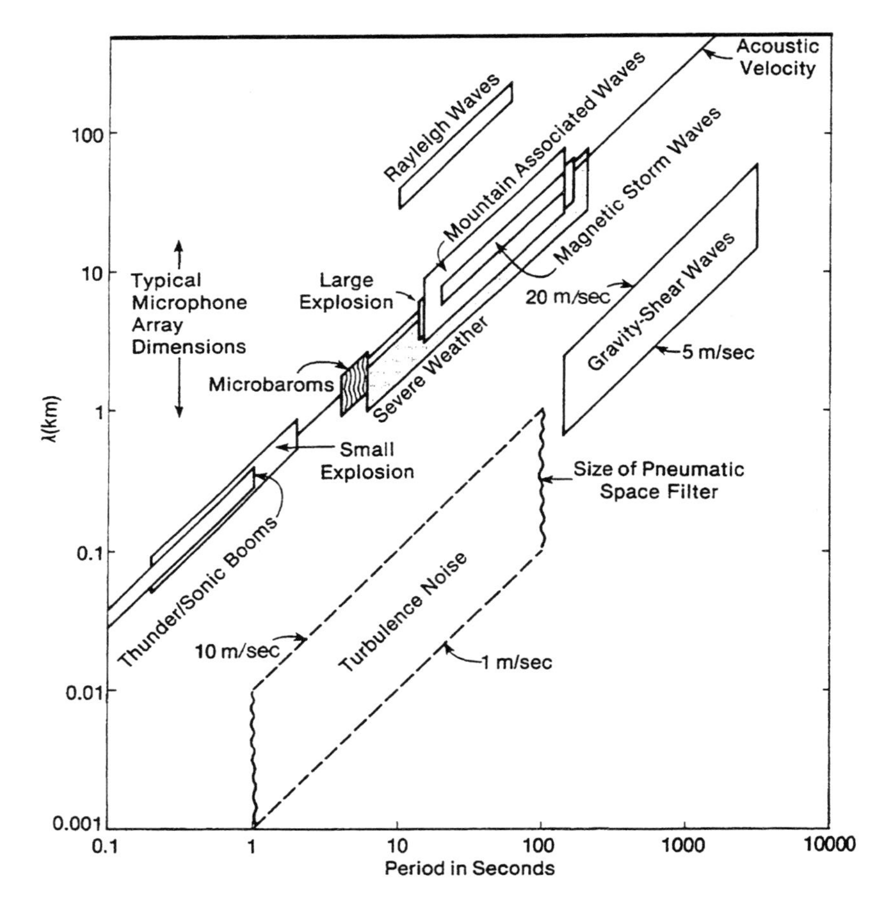

FIGURE F.3 Wavelengths of various types of infrasonic signals. In this and Figure F.1, there is considerable overlap between the properties of explosions and geophysical phenomena.

background sources not of interest in particular types of measurements. One particular background source, especially for southern ocean island stations, is microbaroms with a frequency peak around 0.2 Hz. This part of the spectrum is of interest in explosion monitoring because it is close to the frequency of the main acoustic arrival for 1 kt explosions.

The adopted pressure range curve that the United States presented in Geneva was based on nuclear detections from 0.2 to 112 kt and conventional chemical explosions of less than 1 to 6 kt free air equivalent. These data were given as zero to peak amplitude (scaled by $W^{0.5}$) as a function of distance (where W is yield in kilotons). Although the influence of middle-atmosphere winds on observed amplitude is well documented out to ~2000 km, the data above, in raw form, have a standard deviation of about a factor of 2. This has led some to question whether the wind effects survive to 5000 to 10,000 km. This question requires some additional consideration.

SOUND SOURCES

Natural Sources

There are several natural sources of infrasound that can complicate CTBT monitoring in two ways. First, a background of infrasonic signals from nonexplosive sources can mask the existence of an explosive signal. Second, natural signals similar to explosion signals could cause false alarms or mask the nuclear explosion signal. Thus, there will be great value in documenting the characteristics of various signal types and using this information in the development and application of discrimination algorithms. Table 3.1 indicates a number of natural source types observed by researchers. In the future, additional human-related infrasonic sources may evolve from industrial or transportation methods using high energies. Further research on the range of background infrasonic sources can have significant payoffs in terms of expanded discrimination methodologies.

Explosive Sources

Explosions are impulsive releases of energy within a relatively small source volume. Although the close-in signals from explosions are shock waves, they do contain low-frequency components that can travel to large distances with measurable signal levels. Larger atmospheric nuclear explosions were easily heard at ranges in excess of 10,000 km. Thus, large explosions provide signals well suited to study long-range, low-frequency acoustic propagation in the Earth's atmosphere.

Both nonnuclear events and nuclear explosions are of interest to CTBT monitoring. Nonnuclear sources include explosive volcanic eruptions and bright meteors known as bolides. Infrasonic data from the Mount St. Helens eruption in 1981 can be found in Donn and Balachandran (1981). The volcanic sources may have associated seismicity, which aids in discrimination. Bolides can deposit all of their energy in the atmosphere or survive entry and impact to the ground. In the former case there would not necessarily be another IMS technology detection (e.g., seismic or hydroacoustic), making this an important potential source of false alarms. (There could be ground observer reports or information from satellite sensors.) Current estimates of meteor influx rates give one 15 kt event per year (ReVelle, 1996).

Human-made sources include mining activity, industrial accidents, and sonic booms. Mining operations can use multimillion-pound charges to move large amounts of rock. Many of these events are fired over a finite interval with multiple firing points. These factors reduce the coupling to the atmosphere compared to the same amount of explosive fired as a single source, giving lower amplitudes at a given range. These events do represent human-made sources of great concern because of their size. The CTBT includes guidelines for notification of mining events above a certain size so that they can be identified easily.

Detection of an infrasound signal can confirm the fact that a surface explosion took place, even if no preannouncement occurs. This may be a confidence-building measure. However, seismic and infrasound signals from the event must be compared to address the possibility that signals from a conventional surface explosion have been used to mask a buried nuclear explosion. In cases where seismic and infrasound stations are colocated, the infrasound stations may be useful in identifying signals recorded by seismic instruments that arise because the passage of the infrasound signal causes the Earth's surface to move. Identification of such arrivals can remove signal detections from the data base that must be addressed by the seismic association process and thus increase the effectiveness of that process.

SIGNAL PROPAGATION

The Earth's atmosphere is dynamic, with regions where the wind speed is a significant fraction of the sound speed (the sound speed at the surface is about 340 m/s). This means that the effects of winds must be considered for atmospheric propagation. This is not the case for seismic propagation where the medium is static. The winds of interest for propagation are in the middle and upper atmosphere at about 50 km and 100-120 km altitude. In the middle atmosphere the wind speed can be up to 70 m/s, and in the upper region it can be greater than 100 m/s. In the middle region, the winds are seasonal, blowing east in the winter and west in the summer (in the Northern Hemisphere). During midsummer and midwinter, the magnitude of the zonal component is much larger than the meridional component.

The winds at 50 km couple to the ambient sound speed to form a wind duct at that altitude. For sources on or near the ground, energy moving east in the winter (with wind) can be totally refracted around 50 km altitude and be directed back to the Earth's surface, where, for infrasonic frequencies, it is reflected back into the atmosphere. Total refractions can begin where the sound speed plus the wind speed in the direction of propagation exceeds the sound speed on the ground. Thus, greater wind speeds can form ducts that turn more energy. The process of multiple refraction and subsequent reflection of energy from the Earth's surface is referred to as multi-hop propagation, and the reflections are called bounces. The regions between the first few bounce points are called zones of silence because in a ray picture, no rays—and therefore no sound—falls in them. Ray acoustics is a high-frequency form of the acoustic equations and ignores diffraction effects that can direct sound into the zones of silence. In the opposite direction, against the wind, the only refractions occur at 100 km or higher where the thermal structure alone or with wind refracts the energy

Because of wind dynamics in atmospheric propagation (compared to seismic propagation) the concept of specific phase arrivals is not as useful. Four types of arrivals have been observed and provide a useful distinction. Surface or Lamb (L) waves travel along the Earth's surface and have average group velocities of 330 to 340 m/s. Tropospheric arrivals (T) travel between the surface and troposphere and have travel speeds of 320 to 330 m/s. Stratospheric arrivals (S) travel between the surface and 40 to 60 km and have travel speeds of 280 to 310 m/s. Ionospheric arrivals (I) travel between the surface and 80 to 110 km and have speeds of 220 to 270 m/s. These values are approximate heights of total refraction as given in a ray picture and are reflected in the range of average travel speeds.

With wind propagation, as described above, larger measured signals from a source can be observed downwind than would be measured in the opposite direction, against the wind. This is demonstrated well by Reed (1969) using atmospheric nuclear explosions as sources for infrasonic measurements at several stations around the Nevada Test Site. In addition, average travel speeds are affected by the region through which energy travels.

Dual application of infrasound data for other research goals should be encouraged without distracting from the primary mission of CTBT monitoring. If done wisely, there is potential to enhance the role of the IMS to the benefit of the scientific and technical community. The worldwide, extensive, integrated data sets provided by global infrasound monitoring will constitute a unique resource for monitoring the global environment.

G

Radionuclide Source Term Ranges for Different Test Scenarios*

Radionuclide source terms for a variety of test scenarios determine the percentage of radioactive particulates and noble gases available for global atmospheric transport. DOE experts at Lawrence Livermore and other National Laboratories have considered the various possible uncertainties associated with each source term. There appears to be considerable uncertainty in these data, and research is needed to upgrade them for use in CTBT modeling.

The following is a list of scenarios considered in this study. The range of source terms for gaseous and particulate releases is given in Table G.1.

1. ATMOSPHERIC FREE-AIR TESTS

These are tests conducted in the stratosphere or above the transition zone in the troposphere. The troposphere runs from the Earth's surface up to a range from 10 to 13 km (6 to 8 miles) high where the stratosphere begins. The troposphere is marked by decreasing temperature with height. The stratosphere is a nearly isothermal layer that has its upper boundary at about 50 km (30 miles) above the Earth's surface.

Free-air bursts occur at sufficiently high altitude that no surface debris, soil, or water is drawn up into the fireball. Essentially all of the condensable nuclear debris is in the form of small particulates having radii between 0.01 and 1.0 micrometers (μm). Therefore, little local fallout occurs near the geographic site of the test. The debris deposited in the lower regions forms tropospheric fallout that will reach the surface over a month's time in the general latitude of the test site. The finer particulates deposited in the upper atmosphere form stratospheric fallout that may continue for years with a nearly worldwide distribution.

Radioactive noble gases do not form particulates and will mix and move along with the atmospheric air. Because of the small size of airborne particulates, gases formed by the decay of precursors in the particulates will be able to diffuse out and contribute to the atmospheric noble gas concentration. As shown in Table G.1,

* This material is adapted from a currently unpublished report entitled **"Report of the Peer Review of the Conference on Disarmament International Monitoring System Expert Group (CD/NTB/WP.224 Part II);"** it is referred to in this appendix as "the report."

the source term assumes that almost all of the radioactive particulates and all of the radioactive noble gas are released into the atmosphere.

A. Stratospheric Tests

For high-altitude (stratospheric) tests, it may take several months before any radioactive material can reach the ground. By that time, most of the radionuclides of interest will have decayed. For confirmation of this type of test, stratospheric sampling would be required.

B. Tropospheric Tests

For low-altitude (tropospheric) tests, the debris will quickly reach the surface. Because of radioactive decay, the chemical properties of different fission products, and the fallout rate of different-sized particulates, the radioisotopic composition of the debris will change rapidly. The refractory fission products, including radioisotopes of the lanthanides, rapidly form stable oxides, condensing onto the dust particulates where they fall out more quickly than the more volatile fission fragments. These radioisotopes are useful for providing attribution information. Radionuclides collected far from the explosion are usually highly enriched with the more volatile elements and are only marginally useful for attribution.

C. Tropospheric Tests with Rain-Out

The one significant way to conduct an evasive atmospheric test in a manner that will reduce the

TABLE G.1 Source Term Review

Scenario			Particulates (per cent)	Gases
1.	Atmospheric free-air burst			
	A.	Troposphere, 10-13 km altitude	90-100	100
	B.	Stratosphere, 4-11 km altitude	90-100	100
	C.	+ Rain-out	1-10	1-10[a]
2.	Soil burst			
	A.	Aboveground transition zone, <100 m	20-100	25-100
	B	+ Rain-out	0.2-10	6-15[a]
	C.	Ground surface	15-50	20-50[a]
	D.	Underground transition zone, <70 m	0-15	0.06-15
	E.	Containment	0	0-0.1
3.	Ocean			
	A.	Above-water transition zone, <100 m	50-100	50-100
	B.	+ Rain-out	0.4-10	0.5-10[b]
	C.	Ocean surface	40-60	40-60
	D.	Underwater transition zone, <300 m	1-6	1-40[b]
	E.	Deeply submerged, >300 m	0-1	0.002-1[b]

[a] Debris is assumed to fall or rain out to the soil surface and continually release xenon radioisotopes formed from the decay of their iodine precursors but not from antimony or tellurium precursors.

[b] Values would be considerably higher if a significant part of the xenon dissolved in water or formed by precursor decay in water is able to enter the atmosphere. Releases may occur during conditions of low atmospheric pressure.

likelihood of detection of the radioactive products is to detonate the device during a large and intense rainstorm. Depending on conditions, the amount of airborne particulates will be reduced by a factor of 10 to 100. Although the noble gases will not rain out, the washing out of noble gas precursors such as the isotopes of iodine can reduce the amount of noble gas subsequently released to the atmosphere (Glasstone, 1957). Depending on test conditions, as much as 80 per cent of the nonvolatile fission products may be in the local fallout, including the noble gas precursors, primarily the fission products of tin, antimony, and tellurium.

2. SOIL-BURST TESTS

A. Aboveground Transition Zone Tests

For a 1-kiloton (kt) yield burst exploded in the aboveground transition zone (less than 100 m above the surface of the soil), the strong updraft produced by the explosion will cause large amounts of dirt and debris to be sucked into the atomic fireball and injected into the atmosphere. Depending on test conditions, a large fraction of the fission products will be released into the atmosphere, but much of the particulate matter and some of the noble gas precursors, such as isotopes of tin, antimony, and tellurium, will be removed quickly as local fallout.

B. Aboveground Transition Zone Tests with Rain-Out

Rain-out events, as previously mentioned for tropospheric tests, can affect any of the soil-burst tests that release radioactive fission products promptly into the atmosphere. Again, depending on conditions, the amount of airborne particulates will be reduced by a factor of 10 to 100.

C. Ground Surface Tests

Many of the phenomena and effects of a nuclear explosion occurring on the Earth's surface are similar to those associated with aboveground transition zone airbursts. In surface bursts, however, an even larger amount of rock, soil, and other material will be vaporized and taken up into the fireball. Thus, there will be even larger amounts of local fallout to tie up fission products and keep them from being released.

D. Underground Transition Zone Tests

When a nuclear explosive is detonated under the ground, a fireball is formed consisting of extremely hot gases, including vaporized rock, soil, and bomb residue, at high pressure. If detonation takes place at too shallow a depth (less than about 70 m for a 1 kt burst), the gases will break through the surface and carry up large quantities of rock and debris into the atmosphere. The results are similar to an aboveground burst, but because of the presence of a larger volume of absorbing material, the amounts of particulates and radioactive noble gas are reduced. It is primarily the more volatile species of the fission products that are released into the atmosphere.

For underground nuclear events that are unable to contain gases especially well, xenon-133 may be detected for 25 to 30 days at a distance greater than 300 m and for 30 to 70 days at distances less than 300 m. Thus, there is the potential for use of a mobile noble gas monitoring system to aid in identifying the detonation site.

E. Tests with Containment

The most significant evasive way (from the standpoint of radionuclide monitoring) to conduct an underground test is to detonate it so that there is containment of the hot gases generated by the detonation to keep them from venting at the surface. For a 1 kt explosion a burial depth of greater than about 100 m will contain these gases. Past tests indicate that the burial depth for containment will vary with the cube root of the explosion yield. The most likely release, if any, would be radioactive noble gases. These gases would be released through cracks or fissures penetrating to background sources not of interest to a maximum of the iodine and xenon prompt fission yields and would be released over a period of a few days. Tellurium and antimony precursors formed under these conditions do not readily release their xenon decay products. However, the likely result, using modern published containment practices, is that no gases will be released for a period of weeks to months. By this time, all of the radioactive noble gases except krypton-85 (10.7-year half-life) may have decayed to undetectable levels. At these times, argon-37 (35-day half-life), made from fast neutron transmutation of calcium-40 in the soil, along with krypton-85 may still be detected.

3. OCEAN-BURST TESTS

A. Above-Water Transition Zone Tests

A nuclear device detonated above water in the transition zone will vaporize and carry water up into the fireball. At high altitudes this water will condense to form water droplets, which in turn will form a radioactive cloud similar to ordinary atmospheric clouds. As cooling continues, much of the water, with its suspended radioactive particulates and dissolved fission product ions, will gradually fall back to the surface as rain, spreading radioactivity over a large area of the ocean. It was assumed in this study that no xenon gas would be released from precursor decay in water. This assumption has not been proven.

B. Above-Water Transition Zone Tests with Rain-Out

Rain-out events, as previously described for tropospheric tests, can affect any type of ocean burst test that releases radioactive fission products promptly into the atmosphere. Again, depending on conditions, the amount of airborne particulates will be reduced by a factor of 10 to 100, with particulates being removed in preference to the more volatile fission products.

C. Ocean Surface Tests

When a nuclear device is exploded at or near the surface of the water, the results will be similar to the above-water transition zone tests except that the amount of water drawn into the fireball will increase dramatically.

D. Underwater Transition Zone Tests

Underwater tests may offer one of the best ways to avoid radionuclide detection and/or tion. A device could be detonated and monitored by aircraft and/or surface and underwater vessels that can be long gone before the event can be investigated. In underwater nuclear tests the fireball will be smaller than that formed in airbursts. The resulting bubble of hot gases remains essentially intact until it reaches the surface. At this point, the gases, carrying some liquid and most of the radioactivity, are expelled into the atmosphere. As the pressure of the bubble is released, water rushes into the cavity, forming a hollow column of spray. The radioactive contents of the gas bubble are vented through this hollow column and form a cauliflower-shaped cloud at the top. About 20 seconds after the detonation there will be a massive water fallout that returns much of the radioactivity to the ocean surface. The descending water will form a continuous mass of mist—from the top of the nuclear cloud down to the surface—that eventually is dispersed by the wind. The deeper the point of detonation, the lower will be the amount of radiation released. For a 1 kt device detonated less than about 300 m deep, it is likely that at least some of the volatile fission products will be ejected from the water.

Underwater tests will leave a highly radioactive pool of water above the ocean's thermocline layer. This pool may contain from 25 per cent to nearly 100 per cent of the radioactive debris that disperses slowly compared to atmospheric debris clouds. The thermocline is a thermally stable layer of water that exists in the oceans. A widespread permanent thermocline layer exists beneath the surface layer from depths of about 230 to 900 m. A seasonal thermocline at a much shallower depth forms during summer as a result of solar heating. Thus, it would be important to locate and sample a radioactive pool before it disperses.

E. Deeply Submerged Ocean Tests

Currently no data are available for an underwater test so deep that the gas bubble collapses before reaching the surface. In this case the bubble of hot gases would experience repeated oscillations in both diameter and elevation before collapsing. It appears that a point is reached beyond which the total release of radioactive products does not decrease with depth. Airborne debris from these events consists mostly of radioactive noble gases.

The report also includes the following important conclusions:

1) The particulate source term can be severely reduced or eliminated in many scenarios, especially when rain-out is considered.
2) Gaseous radionuclides are more difficult to conceal than particulates in virtually all scenarios (except possibly, high-altitude events). Thus, a xenon detection system may have the capability to detect nuclear explosions in scenarios where particulate detection fails.
3) In some scenarios, neither gas nor particulate sampling systems will detect an explosion, specifically an underground nuclear explosion that is emplaced using modern containment practices.
4) Remote gas sampling will probably not provide useful attribution information because of the limited information contained in the xenon isotopic signatures.
5) High-quality samples (with minimum fractionation) are the key to attribution. Radionuclides may be especially important for attribution of events detonated over international territory. Aircraft collection of atmospheric debris or collection of water samples from the broad ocean area is necessary to obtain such samples.
6) On-Site Inspection using gas sampling is feasible in the case of moderate leaks but will require precise location information (i.e., distances less than two to three depths of burial) for well-contained events.

H

Acronyms

ACDA	Arms Control and Disarmament Agency
AFOSR	Air Force Office of Scientific Research
AFPL	Air Force Phillips Laboratory
AFTAC	Air Force Technical Applications Center
ARPA	Advanced Research Projects Agency
BAA	Broad Agency Announcement
CD	Conference on Disarmament
CMT	Centroid-Moment Tensor
CSS	Center for Seismic Studies
CTBT	Comprehensive Nuclear Test Ban Treaty
DMC	Data Management Center (IRIS)
DoD	Department of Defense
DOE	Department of Energy
DSWA	Defense Special Weapons Agency (formerly Defense Nuclear Agency)
EMP	electromagnetic pulse
GA	global association
GPS	Global Positioning System
GSETT-3	Group of Scientific Experts Technical Test No. 3
HPGe	high-purity germanium
IDC	International Data Center
IMS	International Monitoring System
IRIS	Incorporated Research Institutions for Seismology
ISC	International Seismological Centre
JVE	Joint Verification Experiment
LASA	Large Aperture Seismic Array
LLD	lower limit of detection
LLNL	Lawrence Livermore National Laboratory
LTA	long-term average

LTBT	Limited Test Ban Treaty
NDC	National Data Center
NEIS	National Earthquake Information Service
NOAA	National Oceanic and Atmospheric Administration
NRC	National Research Council
NSF	National Science Foundation
NTM	national technical means
NTPO	Nuclear Treaty Program Office
NTS	Nevada Test Site
ONR	Office of Naval Research
OSI	On-Site Inspection
PNE	peaceful nuclear explosion
PNNL	Pacific Northwest National Laboratory
PRDA	Program Research Development Announcement
REB	Reviewed Event Bulletin (of the prototype IDC)
SOFAR	Sound Fixing and Ranging
STA	short-term average
SVP	sound velocity profile
TTBT	Threshold Test Ban Treaty
USGS	U.S. Geological Survey